THE
CONSCIOUS
PLANET

A Vision of Sustainability, Peace and Prosperity

How YOU can CHANGE THE WORLD

Neil M. Pine

THE CONCIOUS PLANET
COPYRIGHT © 2023 NEIL PINE

Published by:
Trine Day LLC
PO Box 577
Walterville, OR 97489
1-800-556-2012
www.TrineDay.com
TrineDay@icloud.com

Library of Congress Control Number: 2023952242

Pine, Neil.
–1st ed.
p. cm.

Epub (ISBN-13) 978-1-63424-433-6
Trade Paperback (ISBN-13) 978-1-63424-432-9
1. Veganism. 2. Veganism Environmental aspects. 3. Food habits Environmental aspects United States 4. Vegetarianism Environmental aspects . 5. Agriculture Environmental aspects. 6. Animal industry Environmental aspects . I. Pine, Neil. II. Title

FIRST EDITION
10 9 8 7 6 5 4 3 2 1

Credos is an imprint of TrineDay

Printed in the USA
Distribution to the Trade by:
Independent Publishers Group (IPG)
814 North Franklin Street
Chicago, Illinois 60610
312.337.0747
www.ipgbook.com

Table of Contents

DEDICATION

In Loving Memory of ...

<u>My Father</u>: Norman W Pine, Talented Artist, Craftsmen and retired aircraft worker.

<u>My Sister</u>: Arlene (Kitty) Pine, Published Poet, Artist and Journalist. Graduated Cum Laude from Cal State L.A. and also attended the American University of Cairo in Egypt.

<u>And to My Mother</u> : Florence Pine, a retired accountant, who was also listed in the Publication *"Who's Who of California"* for all her Charitable Work. In her late 90's, she was a *Macrobiotic Vegan*, took no prescription or over the counter drugs and walked 3 miles round trip every day (if weather permitted), to buy her Wheatgrass Juice. She lived till almost 99!

This Book Is Dedicated To Veganism, Sustainability, Health Consciousness, and Peace on Earth

ACKNOWLEDGMENTS

This author would like to thank and acknowledge all the incredible yet very credible sources of information from some of the most brilliant people who inspired him to write this book. Their well-documented and non-bias data was invaluable, and if this book achieves any greatness, it would only be because of all these amazing facts and people that were gleaned by the author in the creation of this literature. It is their research and credibility which gives this publication its true substance and authority.

Mr. Pine would also like to express much gratitude for all the technical help, inspiration and moral support which was afforded to him during the arduous process of creating this book. He sincerely wants to thank everyone who believed in him, as there were doubters and pessimists breathing down his neck every step of the way, including his own family and some people who he thought were his friends!

To start off with, this author wishes to thank Kris Millegan, publisher for TrineDay Press. Mr. Pine feels very proud to be a part of the TrineDay's cutting edge book collection for 2023. Much appreciation also goes to the author's computer technician, Rohon Fullwood. Mr. Pine could not have got this project done without him. An enormous amount of appreciation should be directed towards Vegan television, radio, and media mogul, Mark Thompson, who has endorsed the cover of this book! Mr. Thompson has interviewed Mr. Pine and spoken about his work on KGO radio in San Francisco and KFI radio in Los Angeles. Also, to be thanked is Netflix film maker, Don Ramey Logan, who has provided Mr. Pine, with a letter of intent (LOI), to make a documentary film from this book.

And not to be forgotten, is Mike Russel; publisher, and editor in chief for JetSettingMagazine.com, and Gina Wagner; assistant editor, who have published well over 80 of Mr. Pine's articles. Also to Sydney Murrey, publisher and editor of *Vision Magazine*, for publishing an edited version of chapter "The Insidious Nature of Nuclear Power," in their April 2010 issue, one year before the Fukushima Japanese nuclear disaster! Also, to be thanked, is a very special lady; Carmela Evangelista, of Evangelista Talent Management, who has also officially endorsed this work with her marvelous quote! Much appreciation should also be expressed toward Henry Huang, from the University of California at Riverside (UCR), Sustainability Committee, for allowing Mr. Pine to deliver his powerful speech on Earth Day, 2010, where he received an Eco Hero award which was based on chapter 5 (Peace and Prosperity.) In addition, much thanks to Dr. Zayd Ratansi, N.D. (Managing director of Wellness Without Limits), and to Dr. Carol Sigala PhD, professor of child development, for all their powerful book cover endorsements. Much appreciation also goes to Claudia DeSantis: MS/CNS (Master of Science in Nutrition and Certified Nutritional Specialist, for her all her support and wonderful quote. And finally, to the Lama Jigme Gyatso (Vegan Buddhist monk), who has blessed this book with his divine embellishment!

BIOGRAPHY:

Mr. Pine currently Identifies as a Vegan Zen Buddhist, but grew up in a Jewish family. His grandmother, on his father's side, was the only survivor in her family from the Holocaust! She was very active with charity organizations such as Hadassah and the City of Hope. His mother, Florence Pine, was also very active with these organizations. She was listed in the publication "Who's Who in California" for all her charitable work.

In addition to his concern for animal welfare and the environment, the motivation for Mr Pine to write this book came from a confluence of events which dramatically affected his life! Both his sister and father had passed away from the same form of cancer. He felt helpless and guilty for not being able to help them. His life became dysfunctional, and he was suffering from depression. Just in order to live with himself, he made a commitment to do something great to help the world.

The writing of this book was therapeutic toward his recovery, by creating this work of altruism.

Additional motivation to write this literature also came from his grandfather, Jack Pine, who like John Steinbeck, wrote a novel with a similar theme in symphony with his novel *The Grapes of Wrath*, under the pen name, Paul Rocof. Published in 1938, this book expounds about oppression of Russian peasants during the pre Stalen regime, by a very cruel and powerful Russian oligarchy. *The Rule of the Lash!*[1]

Mr. Pine says that the book is not about him! It's about everything he stands for and believes in! We all need to make money, but Mr. Pine says that when he hears people tell him that his book changed their life, then that means more to him than all the money in the world!

He was writing his first edition of *The Conscious Planet* before Greta Thunberg was even born! His work represents everything she stands for! Veganism is not just a fad! According to *Time* magazine, she was the most influential person in the world in 2019! What people must realize, is that right now, there are literally, hundreds of millions of little Greta's incubating all over the planet! According to the latest polls, vegans, still only represent less than 2 % of the population. Can you imagine the exponential growth potential of veganism and its potential impact on the environment? This book can only become much more significant into the future!

---

1   https://www.amazon.com/Rule-Lash-Paul-Peud-Rocof/dp/B000NW9UD8

PROLOGUE

The writing of this book was both a spiritual and intellectual journey. This is no-holds-barred literature where the author says what he means and means what he says, He does not mince his words! *The Conscious Planet* thinks outside the box and breaks all the rules of traditional thinking. Sometimes this information can be brutally honest or painfully real. This author is not trying to win any popularity contest. He tells it like it really is and doesn't care what anyone thinks!

The Conscious Planet is what the government and big corporations who control it don't want you to know! It is the truth about the truth and nothing but the truth, so help me!

In the modern world, people only know what multinational corporations want them to know, not what they should know! And it is this tainted knowledge, which is creating a society of pathetic drones, who like poor doomed sheep, are too busy just trying to survive to see that the wool is being pulled over their eyes while being led to slaughter! Because these corporate imperatives are being paid for at the expense of the future of mankind!

Before you patronize any product, be conscious about what it's doing to your health, to the earth, and how it will affect future generations. Skeptics have warned the author that this book will receive nominal success, and that it represents a small niche market, where not everyone would be interested in such literature. And in retort to this assertion, it should be acknowledged that yes, it is sadly true, because the information herein may not pertain directly to their life or general wellbeing. So, in understanding who would be interested in this book, let's narrow it down by eliminating the types of individuals who would not directly benefit:

- People who are immortal:
  Talk about good genes; these people really got it made!

- People who have another planet to live on:

This is true; why would they care if the world goes to hell in a hand-basket? See Ya!

Maybe Elon Musk or Jeff Bezos might qualify?

Well, lucky them. However, for the other seven billion people who inhabit the planet earth, who are mortal, and who do not have another planet to live on — then this is absolutely one of the most important books they may ever read!

Become a part of

<u>The Conscious Planet</u>

INTRODUCTION

"The time has come today, there are things to realize"
— (The Chambers Brothers, 1967)

Yes, the time has come today for all of us to realize the responsibility to ourselves, our planet, and to future generations. Open your hearts and minds to the new age of health, sustainability, and compassionate new age concepts in retrospection to the generations of gross polluting, insalubrious, and inhumane practices of the past. Remove the shell around you built up by centuries of anthropogenic ignorance. Strip away the cold-hearted nature of humanity from your being and look deep inside yourself. The truth is there, and the truth is that there is no future in the past! All we can do is learn from our mistakes and evolve.

Whether people realize it or not, eating meat in the modern world represents a cruel and selfish form of decadence. *(No compassion for living creatures, no concern for endangered species and the well-being of almost 1 billion people who are starving on the planet, and no responsibility for patronizing an industry that is the world's leading cause of pandemics, climate change and industrial pollution!)*

"Eating meat is a sickness and raising livestock is the plague!"

Also, in the United States, people are usually only preoccupied with veganism for selfish reasons (health and environmental factors). However, the doctrine of eastern philosophy (AHIMSA: nonviolence and the sacredness of all living creatures), magically transforms veganism into an act of compassion and respect for life. This is a *mystical veneration,* which should be instilled in all children today in order to avoid incidents such as Columbine.

Compassion was the original reason why humans became vegetarians in the first place. Veganism empowers the true altruistic potential of mankind.

It is the realization of oneness with all living creatures and respect for life which will bring together all humanity into the next evolutionary cycle. (Refer to chapter 1 "The Psychology of the Cattle Culture.")

> "It is my view that the vegetarian manner of living, by its purely physical effect on the human temperament, would most beneficially influence the lot of mankind."
>
> – (Albert Einstein)

If we have truly evolved to the point where we no longer must inflict pain and suffering upon other living creatures, in order to survive, then why must we continue to do so? *"Does something have to die in order for you to live?"* *The Conscious Planet* questions people's typical concept of reality, examining the traditions and ethos which have led to the moral justification of animal slaughter. *The Conscious Planet* also questions the word *civilization*. What is civil? What is humane?

> *"Kindness and compassion towards all living things is a mark of a civilized society."*
>
> – (Cesar Chavez)

Americans are "in-fat-you-ate-it" with beef! On average, Americans consume an equivalent of seven 1,100-pound steers in their lifetime. However, what the cattle industry is not telling them is that statistically, it takes 2,500 gallons of water to produce just one pound of beef, as opposed to only 6 gallons of water to produce a pound of broccoli. This has been a contingent factor of drought in many parts of the world. Due to climate change and burgeoning third world populations, wars are expected to arise in the future due to shortages of water! Will water become the new oil? (See chapter 7 "Dust Drought and Desertification.")

We really can change the world just by making certain lifestyle choices that really are not a sacrifice at all. The only reason that people believe that they are making a sacrifice is due to corporate *subreption* and propaganda, taking advantage of people's weaknesses and inhibitions. Through media hype, the livestock industry wants to keep the public *environmentally, physically* and *spiritually* unconscious so that people will continue to stuff their face with all their *bad karma*!

Also, with dramatically rising commodity prices (gasoline, real estate, food, etc.), people are still led to believe that the meat they buy is affordable. What they're not aware of is the tens of billions of dollars earmarked by the Environmental Protection Agency (EPA) to clean up water pollution, improve soil degradation, and deal with desertification from drought, all a contingent factor of livestock production. The U.S. taxpayers will eventually be forced to flip the bill for all this profligate environmental negligence! *"The time bomb is ticking!"* (Refer to chapter "Pollution: Cattle, Petrochemicals, Heavy Metals, Fluorides, Dioxins, and PCBs")

Anthropology dictates that human nature fears change by clinging on to the past as a security blanket. After hundreds of years of tradition, many people's mind-sets still remain unfazed. Time and evolution will inevitably bring change! However, it is of paramount importance in the realization of this element before it's too late! By making a conscious effort to abstain from animal products, we help to abolish global warming, pollution, famine, and the oppression of third world people!

Naïve opinion would go as far as to label the *Vegan Anarchic* principles of *The Conscious Planet* as heresy (an opinion or doctrine contrary to generally accepted beliefs, religious opinion contrary to established church dogma). To criticize government policy and religious dogma carries unpatriotic, heretical, and even blasphemous implications. However, the truth is that we are not just talking about man's next great step in evolution, but also about something even more critically important: his survival on the planet earth! *Man must take this evolutionary step in order to just survive!*

We must caste aside our antediluvian traditions and ethos and maintain a salient view regarding corporate imperatives (who only want to get us hooked on animal products, junk foods, and drugs). It's all about improving our health, being humane, respecting the environment and really caring about future generations. *Now how unpatriotic, heretical, and blasphemous could that be?* If man does not make this critical paradigm shift, then credible scientific documentation suggests that the world will face cataclysmic ecological destruction! To reach this goal, and therefore circumvent the process of pollution, climate change, and endangered species, we must live green and sustainably.

"If we wish to make a change in the world, then we mu we must become part of that change"

– (Gandhi)

This is way beyond any self-help book you have ever read. Not only does this information help a person psychologically, but it will also dramatically improve them physically, and in turn, will ameliorate all external environmental factors surrounding that individual, making the world a better place to live for everybody.

The Conscious Planet is the ecological, physiological, and compassionate psychological justification for sustainability which entails an organic vegan macrobiotic diet. In addition, this literature represents a conglomeration of information that major corporate and government entities *don't want you to know!* This obfuscation of the truth is based on consumerism, militarism, media hype, and corporate imperatives which instill the standards and concepts that form our modern Western culture.

Politicians work for the corporations through lobbyists who grease their palms so that their corporate infrastructure will remain inexpugnable. *"Politicians are merely instruments of corporate proclivity."* These corporate entities that rule the world have no heart or soul, only their bottom-line imperatives to make as much money as possible, while still trying to present a facade of integrity to the American public. Case in point, the Altria Corporation, formerly Phillip Morris, who publicly portray themselves as a benevolent organization rather than the most significant factor in an industry that is responsible for more than 150 times the American deaths annually, than Osama Bin Laden's attack on the World Trade Center! The same is also true of the American Medical Association (AMA) along with the National Beef and Dairy Councils, which are controlled by the insidious pharmaceutical and livestock industries, respectively. All these multinational corporations, which control our politicians and the rest of the world, do not necessarily have the United States or its citizens in their best interest.

The Conscious Planet is just plain common sense for the new millennium. When your child runs a temperature, you do everything in your power to bring it down. Well, our Mother Earth is running a serious temperature, and we must all get together and synchronize our efforts in order to cool her down, or we all face cataclysmic environmental repercussions! We must put aside our antiquated notions of excess. However, this author points out that with technology *"we can still have our vegan cake and eat it too."*

It is the goal of this publication to enable man to survive into the twenty-second century. However, the way our current global political agenda and multinational corporate infrastructure is evolving, it appears that we are headed toward self-destruction! We need a radical change in awareness, not just in the United States, but on a worldwide basis!

The word strutheous is an excellent word meaning to be like an ostrich and place your head in a hole to avoid problems, which is an impractical behavior. However, this mind set is so typical of our Western society today. So, it was not at all surprising, while this author attempted to explain some of the alarming and sobering concepts behind *The Conscious Planet*: (GMOs, Radiation, Vaccines, Pink Slime, etc.), to his next-door neighbor, that the man stated *"I really don't want to know."* Doesn't this attitude clearly exhibit the strutheous nature of today's modern western culture? To blindly go along with all the "bovine excrement" of today's MSM (Main Stream Media). This credulous mentality by the average American is what the major corporations are banking on!

Furthermore, *The Conscious Planet* embellishes the reader with fascinating information about endangered species' (fauna and flora), cause and effect, including a section just on endangered insects. In all corners of the globe, cattle are wreaking havoc over the environment, heating up the planet with carbon dioxide and methane gas, melting the polar ice caps, while creating pandemics, drought, famine, and endangered species.

The author also enlightens the reader as to the latest innovations in alternative energy and transportation technologies, while also exposing the corruption and outrageous environmental policies by the Trump, Bush and Reagan administrations, who were all funneling tax dollars to the *fat cat* polluters and purveyors of dangerous technologies (oil, coal, and nuclear), rather than to give tax credits to develop safe alternatives. (See chapters 3, "Peace and Prosperity," 4, "Veganomics" and 16, "The Insidious Nature of Nuclear Power") An edited version of this antinuclear chapter was published in *Vision* magazine, April 2010 issue (One year before Fukushima!)

Also, the pseudo-science of genetically modified organisms (GMOs), is examined, (see chapter 12 "Implications of Genetically Modified Organisms"). Even though technically vegan, many foods have become victim to the junk and fast-food industries insidious protocol. Veganism should mean more than just merely not using animal products; it should also mean respect for life and purity.

The Conscious Planet is one of the most profound, and controversial pieces of literature that you may ever read! It is impeccably researched, using brutally honest and shockingly true facts, along with powerful conjecture, to drive home the point of health, sustainability, compassion, and peace on earth! If you sincerely read this literature with an open heart and an open mind, then it will change your life forever, the way you think of yourself and look at the world around you.

Special Note:

The original version of this book was so comprehensive (560 pages), that the publisher said that it should be split into 2 books: The new version of *The Conscious Planet*, will now mainly be focusing on vegan psychology and environmental aspects, and book 2, *The State of the Art Lifestyle*, will mainly deal with health issues and cutting edge breakthroughs in natural living.

However, because many health issues are related to ecology, then several chapters may overlap in each book.

Some brilliant key elements prognosticated in Mr. Pine's first edition of *The Conscious Planet* (Originally written over 15 years ago), are more significant now than at any other time in history! Global pandemics, Climate change, the Rainforests burning down, the dangers of GMOs, problems with opioids and prescription psyche drugs, Superbugs, the benefits of medical marijuana, Fukushima radiation and much more!

Chapters dealing with, weight loss, thyroid health and protection, cancer prevention and holistic treatments, raw juicing, vegan macrobiotic system, the benefits of medical marijuana along with the negative impacts of substance abuse, prescription drugs, microwave ovens, artificial sweeteners, animal products, and electromagnetic interference, and much more, are all covered in Mr. Pine's second book *The State of the Art Lifestyle*.

It is this author's sincere hope that he may
enlighten the public as to the altruistic vision of
this anthropogenic paradigm …
known as…

The Conscious Planet

Zen Principles

I am nobody...yet I am everybody

I am the purr of a kitten the roar of the tiger

and the glint in an eagle's eye

I am nowhere...yet I am everywhere

I am the grass growing up through the earth the light of the sun

and the rain falling down

I am nothing..................I know nothing................and think of nothing

yet I realize everything

I am only a mere mortal ....yet my spirit will live on for eternity.

– (Neil M. Pine)

TheConsciousPlanet.org

Chapter 1

What is Veganism?

Veganism is one of the hippest and fastest growing lifestyles, which is taking the world by storm, especially in light of the growing trend in health, environmental awareness, and compassionate psychology.

But what does it really mean?

Over the years, the word vegan has taken on an ambiguous definition. Most people consider anyone who uses no animal products to be a vegan, but there's so more to it than that. According to Wikipedia, veganism is "the practice of abstaining from the use of animal products, particularly in diet, as well as an associated philosophy that rejects the commodity status of sentient animals. A follower of veganism is known as a vegan."

The contingent factors of a vegan philosophy are animal rights and compassion for all living creatures. A true vegan philosophy will not tolerate any products which may harm or exploit animals. Vegans will not wear leather, fur, wool, silk, or use bee products such as honey, pollen, or royal jelly.

A person who eats no animal products for health or environmental concerns, but still may engage in hunting, fishing or patronizing products which may exploit animals, are technically referred to as being on a *"Plant Based"* diet.

The word vegetarian is defined by people who are primarily plant based, and traditionally do not intend to harm animals, but may still exploit them through the ingestion of their bodily secretions and embryos. Lacto-ovo-vegetarians: (Milk, eggs, cheese, butter, mayonnaise, honey, etc.) Also, like vegans, most vegetarians are dogmatically opposed to wearing fur or leather, but unlike vegans, they may still justify wearing wool or silk.

Also, in the United States, people are usually only preoccupied with vegetarianism for selfish reasons (health and environmental factors). However, the doctrine of eastern philosophy (AHIMSA: nonviolence and the sacredness of all living creatures), magically transforms vegetarianism into an act of compassion and respect for life. This is a mystical veneration which should be instilled in all children today in order to avoid incidents such as Columbine. All mass murderers and serial killers abused animals before they ever thought about hurting people!

Compassion was the original reason why man became a vegetarian in the first place!

However, in the modern world, we now realize the inhumane and profligate environmental repercussions of even raising dairy cattle. Today, any vegetarian who uses dairy products, but still claim that they are vegetarian for compassionate reasons, are just fooling themselves! (The majority of Buddhists and Hindus in the world fall under this category) A dairy cow can live up to 25 years, but under the heavy burden of modern industrial milk production, they will only live for four years, and then get ground into hamburger meat. Also, the baby calves of dairy cows are heartlessly separated from their mothers at birth and most of the time killed if not female!

> *"Veganism empowers the true altruistic potential of mankind. It is the realization of oneness with living creatures and respect for life and the environment, which will bring together all humanity to the next stage of evolution!"*
>
> – (Neil M. Pine)

CHAPTER 2

THE PSYCHOLOGY
OF THE CATTLE CULTURE

The ubiquitous saturation of animal products through media advertisement is a major part of the socioeconomic structure of our modern western civilization. While the public remains in a somnambulistic state of consumerism, the insidious forces of the modern industrial livestock complex perpetrate their very psyche, by media hype, through traditional ethos, which create the accepted concepts of a utilitarian society.

Throughout history, cattle have always had a significant impact in the development of man's sociology and religion. Cattle were worshipped from the earliest recorded history of man. These hearty ungulates were worshipped in celebration of their fecundity, virility, and utilitarian value. [1]

Today, many millennia later, culinary shrines have been erected for worship in the visceral tradition of barbequing! When it comes to traditional foods, billions of people from around the world emulate their forefather's dietary customs. These traditional foods also seem to play an important role in man's cultural identity.

The dietary habits handed down from generations seems innocuous at first glance, but when enough people look beneath the surface and start to realize the negative impact that cattle are having on our ecosystems, then the earth may finally reach a critical turning point!

There is nothing wrong with tradition; however, when it starts to circumvent the evolution and therefore, the survival of mankind, then we must put all our antiquated dogma and traditions aside and work toward a common goal of sustainability.

Many people feel that by the deprivation of meat and dairy products, they would be denying the American dream itself; that the use of adipose products are symbols of affluence and prosperity in our Western culture, and by the abstinence of these foods, they would be contradicting everything they and their forefathers had worked for.

During the early part of the twentieth century, eating a steak every day was considered to be a status symbol, the mark of success, and the equivalent to driving a luxury car. And due to modern day corporate *"Subreption,"* meat is still a way of life for the average American, and it is this mentality which spells out imminent peril for our ecosystems, endangered species, and the health of the American public![2] & [3]

Our forbearance was based on an ideal which primarily involved survival and overcoming adversity, while forging their way into the new world. John Winthrop, early American politician and religious leader, viewed the American wilderness as *"a force of evil to be overcome by God's glory."* [4]

Back in those days, with abundant resources, people were not concerned about conservation, and probably could not have fathomed what a significant role that it would play in the future. But now we realize what sustainability means, and to patronize the cattle industry in light of egregious environmental evidence seems ludicrous!

We still live in a *"meat and potatoes"* society, from John Wayne to McDonald's, all representations of the American cattle culture, where brute strength and size become sex symbols, a sign of *"machismo,"* as millions of children and teenagers scramble to emulate their favorite sports heroes at everything from athletic shoes to eating burgers. Meat represents strength and virility; it's more American than *"apple pie"*!

Consequently, there should be no surprise that mega conglomerate burger chains have tried to capitalize on this macho patriarchal image through the mainstream media (MSM), by using factitious clichés or innuendos, which insinuate that men who don't eat meat are *"wimps, sissies, weak, pathetic, impotent, unpatriotic, kooks, oddballs, tree-huggers, or not real men,"* rather than to be praised for the *"health conscious," "compassionate,"* and *"ecologically chivalrous"* individuals to which they really are!

A prime example of this was a Jack-in-the-Box commercial which aired in 2007. In this commercial, Jack is sitting in what appears to be an elementary school auditorium, while his supposed son delivers a speech on stage about life goals. He states *"When I grow up I want to be a vegetarian."* What a noble statement, enough to make any parent proud who understands the dynamics of health and sustainability in regards to dietary practice. Instead however, upon this deliberation, Jack hangs his *"overstuffed, artery-clogged head"* in woe, as if you should be ashamed of your child for wanting to save the planet and practice compassion for animals?

Directly thereafter, the boy corrects himself and says, *"I meant veterinarian."* The camera then pans back to Jack who is now standing up and clapping enthusiastically. [5]

Because these huge corporations like Jack-in-the-Box, McDonald's and Burger King, pay for TV, radio, and newspaper ads, people are never warned, through these traditional media sources, that the worldwide production of livestock is responsible for more global warming gasses than that of fossil fuels, besides being the leading causes of drought, famine, pollution and rainforest destruction! [6]

Today's white-collar executives of these giant conglomerate meat producers, almost never see or experience the cruel, heartless, morbid and bloody nature of their business. All they see are numbers with dollar signs, compassion being completely left out of the equation! These corporations rape and pillage the earth while at the same time exploiting minimum wage zombies, who do their dirty work by serving up heaping helpings of tortured, chemically imbued, rotting flesh!

During the Middle Ages and preindustrial eras, butchers were regarded with repugnance and disdain. Because of their cruel disposition, butchers were barred from jury duty in cases involving capital punishment. In fact, in the Oxford English Dictionary reference, published in 1657, butchers were depicted as *"greasy, bloody, slaughtering, merciless, heartless, pitiless, crude, rude, grim, harsh, stern, and surly."* [7]

In today's modern world, all sacredness of life has been replaced by inhumane protocol. As compassionate and sustainable author Jeremy Rifkin puts it, there exists *"a coldhearted evil,"* which permeates throughout our modern Western civilization!

All the *horror*, the *fear*, the *shame*, the *guilt* and the *repulsion* of animal slaughter, are hidden behind a facade of impersonalization and marketing glitz! Heartless corporate imperatives, expedited commodities, flashy advertising, fancy packaging; neat little units of sustenance, available twenty-four hours a day, no thought ever given to the suffering incurred by other living creatures!

"The modern industrial livestock complex wants to keep the public environmentally, physically, and spiritually unconscious, so that people will continue to stuff their face with all that bad karma!"

– (Neil M. Pine)

NEGATIVE ENVIRONMENTAL FACTORS ASSOCIATED WITH LIVESTOCK PRODUCTION

Livestock production is a contingent factor in every ecological catastrophe we face; pandemics, pollution, famine, endangered species, rainforest destruction, wildfires, drought, desertification, and antibiotic strains of infectious diseases! (Superbugs)

• **Number 1 Cause of Famine:** Almost one billion people are starving on the planet. Up to 60 million people per year die of starvation or diseases related to malnutrition, (mostly children), while at the same time enough grain is fed to livestock *"every day"* to feed approx. *"Fifty"* billion people! [8]

(See chapter 8, "Overfed Cattle Equals Starving People.")

• **Number 1 Cause of Global Warming:** Methane, CO2, ammonia, along with other effluvium emissions related to the cattle industry. In a report issued by the World Watch Institute, in 2018, they have dramatically revised the aforementioned U.N. report, to reflect that livestock are responsible for 51% of all global greenhouse gas production! [9], [10] (See chapter 6: "Cattle and Egregious Greenhouse Gasses!"

• **Number 1 Cause of Pandemics:** Except for Covid, which was most likely man made, virtually all pandemics throughout history have been linked to animal agriculture. (See chapter "The Malicious Mechanics of Modern Meat"")

• **Number 1 Cause of Drought, Desertification, and Rain Forest Destruction:** Cattle are undeniably wreaking havoc on all six continents of the planet! Statistically, it takes 2500 gallons of water to produce just one pound of beef as opposed to only 6 gallons for a pound of broccoli. 80,000 acres of rainforest are destroyed every day, primarily due to livestock production. This is what inspired the book cover. [11]

(Refer to chapter 7 "Dust, Drought, and Desertification.")

• **Number 1 Cause of Endangered Species:** Man's greed and lust for beef is driving wildlife into extinction as natural habitats are destroyed in order to create a relentless proliferation of livestock production. The lastest U.N. report states that one million animal and plant species now face extinction! Ecosystems and natural biomes are degrading at a rate unprecedented throughout history! However, hundreds of new species, including plant life, are being threatened every year. Therefore, we need to bring more awareness to this problem! According to many credible scientists: *"We are experiencing the worlds' sixth mass extinction!"* [12]

(See Endangered Species section)

• **Number 1 Cause of Pollution: (Ground, air and water):** Twice that of all industrial pollution combined! Livestock are responsible for the production of over 1.2 trillion tons of manure annually; polluting lakes, rivers, underground water tables and creating dead zones in the ocean from all the runoff! [13]

(See chapter 7, "Pollution: Livestock, Fossil Fuels, Heavy Metals, Fluorides, Dioxins, and PCBs.")

Negative Health Factors Associated With Livestock Production & Meat Consumption:

- Heart Disease
- Intestinal Disorders (Acid Reflux, Crones Disease/Ulcers)
- Breast, Colon, Rectal, and Stomach Cancers
- Alzheimer's, Dementia, Senility
- Diabetes
- Obesity
- Gout
- Kidney and Liver Damage
- Osteoporosis
- Pancreatic cancer
- Prostate cancer
- Ovarian cancer
- Endometrial cancer

- Breast cancer
- Kidney stones
- Cervical cancer
- Hyperglycemia
- Constipation
- Hemorrhoids
- Hiatal hernias
- Diverticulosis
- Gallstones
- Hypertension
- Asthma
- Salmonellosis
- Trichinosis
- Gout
- Irritable colon syndrome
- Creutzfeld-Jakob disease (variant of Mad cow disease)
- Flu viruses
- E-coli
- Mumps
- Measles
- Smallpox

Source: Vegetarian Society [14]

"Eating meat is a sickness and raising livestock is the plague!"

– (Neil M Pine)

TEXAS TIRADE

We are indeed a cattle culture from the early days of our forefathers, and this cowboy image and romanticism is still entwined in the hubris of the American public in the form books, cinema, television, and magazines: the Marlboro man herding cattle, Black Angus commercials, Chaps cologne, Carl's Western Bacon Cheeseburger, and Louis Lamour, with over one quarter billion books in print.

Popular Western author, Emerson Hough, characterizes the epitome of the aggressive and jingoistic nature of our Western cattle culture: *"The beef herders and beef eaters of history have been winning peoples, not the vegetarian nations."* [15]

The *"rough and tumble," "roughhoused," "roughneck," "roughrider"* type of individual (the cowboy), is one that is characterized by the *"carnivore,"* who eats his meat rare and is a *"real man."* Besides the use of sugar, there is a definite correlation between eating large quantities of meat and *reprobative* antisocial behavior. The overabundance of hormones and adrenaline in these products have historically been associated with much overaggressive, deviant, and irascible behavior!

The International Vegetarian Union (IVU), in their report entitled *"The Effect of Animal Protein on Human Behavior,"* they refer to a calcium and phosphorous imbalance, along with certain deposits that form on the brain created by the consumption of meat, which may lead to personal belligerence. Has the American *"cattle culture"* spawned the creation of *"modern-day troglodytes?"* [16]

Traditionally, the cowboy represents an archetypal image of the *"all-American hero,"* (John Wayne, Roy Rogers, Gene Autry, etc.). When you picture this stereotypical image, the most prevalent state of the United States which comes to mind is Texas; the Lone Star State.

Historically, Texas epitomizes the "Wild West;" guns, outlaws, shootouts, brothels, red meat, cattle ranching, and oil wells—everything uncivilized, illicit, non-compassionate, unsustainable, and gross polluting!

This is the same state where one of the nation's worst sniper attacks and presidential assassinations took place: the University of Austin [17] and JFK [18], respectively.

Texas is the only state where, historically, a man could get away with killing his wife for infidelity; yet if it were the other way around, the wife would get the death penalty or go to prison. In Texas, an archaic law stands, but is no longer legally enforced, which states that if a man catches his spouse in mid coitus while on his own property, then he legally has the right to kill, not only his wife (the adulteress), but her partner as well (the adulterer). Crime of passion is a defense used as a mitigating factor by defendants in order to prove blind provocation, or an *"at the spur of the moment"* judgment, with no abatement period, in order to have time to think about or premeditate the crime [19]

Traditionally, women don't get a lot of respect in Texas. The "Lecherous Lone Star State" has a reputation synonymous with strip bars and prostitution. Most people don't know that Dolly Parton is from Tennessee, but everybody knows where *"The Best Little Whorehouse"* is from! And not surprisingly, Texas ranks among the top seven states in terms of sex offences, with 2,270 offenders per capita, to every one million people as of 2010. [20]

In addition, Texas is also one of the only states which has ever attempted to secede from the union. Many ethnocentric [21] Texans today still support such a secession. This is the actual website which supports this succession. http://www.anus.com/etc/texas. Anus dot com? This author didn't make this up! Does this mean that you have to be some kind of a *Tex-A-Hole* to believe that a bunch of *"reprobative rednecks"* could ever be responsible enough to run their own country?

This website spews such rhetoric as *"Washington, D.C., is not just many miles from Texas, but it's far from the heart and soul that makes Texas great."* Anthropogenically speaking, what makes Texas great, is also a *"great embarrassment"* to the future of sustainability and the evolution of mankind! [22]

They talk about heart and soul? What heart and soul? It's more like the heartless soul of a bunch of slaughtering, womanizing, rowdy, gross polluting, *"non-bio-sustainable barbarians,"* who will leave behind a legacy of violence, inegalitarian and chauvinistic ideals, along with uncompassionate and destructive environmental practices! Didn't we already learn our lesson the last time we let someone from Texas run this country?

Now who would have ever guessed that our kind and benevolent Lone Star State also holds the record, on an annual basis, for the most executions ever committed? And speaking of barbarians, George W. Bush, "our modern-day Pontius Pilate" [23], during his five-year term as governor of Texas, was appropriately nicknamed the *"Texecutioner."* He personally helped to make *"Texas, execution capital of the modern world."* [24]

One hundred and thirty-one prisoners were heartlessly executed under the pharisaical whims of Governor George W. Bush. Most of these prisoners were poor or indigent, and therefore, could not afford competent legal counsel, and due to a corrupt and draconian Texas judicial system, were not afforded the proper constitutional rights in order to conclude such serious determinations against them in regard to capital punishment. Of these 131 doomed prisoners, 9 were later proven to be innocent, and who knows how many more were innocent, which could not be proven? No wonder Texas wants to secede from the Union; they run their state like a third world nation! [25]

Was G. W. Bush unusually cruel? Most definitely, however, doesn't this statistic also give further merit to the degenerative and antisocial nature of this scathing Lone Star diatribe? As of 2009, almost a decade after Bush's tenure in office as governor, Texas still held the record for annual number of executions with twenty-four, which represented almost half of all the executions in the United States. Alabama was second with only six executions during the same time period. [26]

Saddam Hussein was executed by a puppet government installed by George H.W. Bush. It was also proven that the Bush administration falsified evidence in order to justify the invasion of Iraq. [27] Therefore, George Bush was originally elected in the same way that Saddam and many of Bush's fellow Texans were condemned: "under false pretenses!"

1/28/10: A special *60 Minutes* report revealed that now-retired Supreme Court Justice, John Paul Stevens, stated that he believes that throughout his career, one of the greatest constitutional injustices in history was the primary election of 2000, where Al Gore was cheated out of the vote by house majority rule, which mandated that a recount would not be necessary. Yet at that time, Stevens expostulated his understanding to the Senate Judiciary Committee that there was no constitutional basis in order to justify house majority rule! He saw no reason to not demand a recount and felt that something was terribly wrong with the system at that time. To this day, he still feels very guilty about the outcome, stating *"The entire history of the world would have been different, if we only could have initiated that recount!"* (Refer to chapter 3 "Peace and Prosperity.") [28]

This backward, uncompassionate, and socially oppressive culture is further exploited by the sociopathic killers in the semi true story of *The Texas Chain Saw Massacre*. According to the facts surrounding this plot, the perpetrators (Leather-face and family), were disgruntled slaughterhouse workers due to being laid off on account of the region of Texas where they live becoming economically disenfranchised by exposure to toxic waste from the petrochemical industry. Many of their relatives became terminally ill, and offspring deformed. Therefore, their opprobrious behavior was motivated through their desire to get even with society by treating other humans the way that they were taught to treat cattle! This film primarily revolves around violence against women. The director of this movie, Tobe Hooper, cites cultural and political changes affecting our modern Western culture. [29]

According to statistics, virtually all serial killers and mass murderers initiated their egregious behavior by torturing, mutilating, and/or killing animals when they were young. The root of this sociopathic enigma stems from the word **Speciesism:** *An assumption of human superiority over animals based on Darwinist (Primarily Judeo Christian and Moslem), Anthropocentric beliefs.* [30]

> Anthropocentrism: (a) Assuming that one's faith takes precedence over all other beliefs. (b): A theory in the interpretation of the world as in terms of Human values or experiences. (c): An assumption based on religious doctrine, that human beings are the most significant entity in the universe. [31]

Speciesist phraseology has become engrained in the psyche of our modern Western civilization. But we may not realize the deep negative psychological impact that these figures of speech subconsciously send to our children, e.g., *"there's more than one way to skin a cat"* and *"to kill two birds with one stone."* Theoretically, these analogies could be made while a child is problem solving. However, if you tell a young child that *"there's more than one way to skin a cat,"* then this statement could conjure up very disturbing images in a child's mind. To skin a cat? Isn't this considered a cruel and inhumane act in modern society? And for the sake of efficiency, to teach children the saying *"to kill two birds with one stone"* also displays the same callous nature of our Western culture. Young children instinctively love animals; they would never think of harming them. We, as adults, introduce the idea of slaughter to our children. After such Speciesist comments, a child might think, *"Why would someone want to skin a cute cat like mine or kill the pretty little birds that sing to me in the morning?"*

Research scientists, in conjunction with the FBI and other law enforcement agencies nationwide, have conclusively linked animal abuse to domestic violence, rape, child abuse, serial killing, and to the recent rash of school related killings by adolescents, according to Dr. Randall Lockwood, vice president of Training Initiative for the Humane Society of the United States (HSUS). [32]

In a related study conducted by the Massachusetts Society for the Prevention of Cruelty to Animals (SPCA), animal abusers have a 500% greater propensity of committing violence against humans. Also, spousal abusers tend to be big meat eaters. This makes perfect sense, since eating meat is abusive toward any animal who is murdered for its flesh. *Think about it, it's only people who eat meat who would ever want to abuse animals or people in the first place!* The **FBI** has recognized this connection between animal abuse and violent crimes since the early 1970s when it was determined, while analyzing the lives of serial killers, that virtually all of them demonstrated *"childhood and adolescent histories of serious and repeated animal cruelty."* [33]

The American Psychiatric Association also concurs that *"a pattern of violent acts against humans directly stems from violence against animals."*

Therefore, according to the American Psychiatric Association:

> "Abusing an animal is a way for a human to find power/joy/fulfillment through the torture of a victim they know cannot defend itself."
>
> "Rape is a way for a human to find power/joy/fulfillment through the torture of a victim they know cannot defend themselves. Child abuse is a way for a human to find power/joy/fulfillment through the torture of a victim they know cannot defend themselves." [34]

What the American Psychiatric Association failed to mention is that the mere act of eating meat is also without question *"a way for a human to find power/joy/fulfillment through the torture of a victim they know cannot defend itself."*

Many people consider themselves animal lovers, but think nothing of eating a juicy steak or piece of pork. What do you tell your children when they come home from the petting zoo and you're making spare ribs? And they ask the question, *"Isn't this a pig like the one at the zoo?"* (By the way, pot-bellied pigs make excellent pets for children.) Now you can try to cover it up by quoting your religious dogma, but your child may still remain confused the rest of their life!

In the story of The Silence of the Lambs, it gives a similar twist to this assertion. The heroine of the story, which is played by Jody Foster in the movie, makes an analogy of lambs being slaughtered to that of cannibalism. As an orphaned girl, growing up on a Midwest farm, she developed a deep relationship with one of these gentile animals, akin to being her brother or sister. But realizing that the lamb will be slaughtered, she makes a desperate failing attempt to save its life. And when she realizes what her family is preparing for dinner, this makes her feel like a cannibal, because quite literally, how could you eat a member of your own family? This analogy relates to the movie in that the woman victim of the story was captive of the notorious cannibal and serial killer, "Buffalo Bill," and Jodie Foster feels an affinity toward the victim like *"a lamb waiting to be slaughtered!"* [35]

By introducing vegan family values, we bring our children to the realization of love and respect for all living creatures. This magical relationship between man and beast is one of the most precious qualities of life, a mystical veneration which should be instilled in all children, to teach compassion, thus avoiding future *"Columbinish"* scenarios. By eating meat, we automatically patronize slaughter, and therefore we can only impugn this mystical relationship!

A Good Question: Is there, or has there ever been, a *"vegan serial killer"*? Technically speaking, using the same verbiage while completely discounting lexical semantics, the answer is yes! (seer.ial) killer? This author wishes to confess to the millions of readers about his heinous acts of gluttonous cruelty against all grains on earth! He freely admits to heartlessly *"putting away"* hundreds of bowls of granola. So let's get the facts straight:

Vegan Serial Killers: Zero!
Vegan Cereal Killers: Millions and growing!

And while on the subject of mass murderers, violence, and Texas, once again, our favorite Texan, G. W. Bush, was accused of having knowledge or involvement with the ritualistic killings of seventeen people, including many children, at Brownsville, Texas, in 1984. One of the ink pens from Bush's oil company was used to gouge out a victim's eye! At the gory crime scene were the words scrawled in bloody two-foot-tall capitalized letters: *A CHARGE TO KEEP.* [36]

28

Bush claims that he was out of town at the time of the killings and had no knowledge of this incident. Yet sixteen years later, he published a book entitled *A Charge To Keep: My Journey to the White House,* coincidentally, all spelled with uppercase letters. [37]

In modern conventional legalese, isn't this called a "Confession," or at least an acknowledgment of guilt which links Bush to this shocking crime scene? What gall! Doesn't this display pompous audacity for a presidential candidate to name his book, regarding his incumbency, after a gruesome and tragic mass murder scene?

Not only was Bush affiliated with this satanic cult, but he also pardoned (Henry Lee Lucas), one of the most notorious mass murderers who just also happened to be tied to these same ritualistic killings! In Bush's entire tenure as Governor, he had never given clemency to any other condemned prisoners, even after evangelist Pat Robertson pleaded for a woman (Betty Lou Beets), and also executing a great grandmother (Karla Fay Tucker), who was convicted of killing her chronically abusive husband! But all of a sudden, he realizes empathy near the end of his tenure, and after 130 unabated executions, he decides to pardon one of the world's most heinous killers?

The crimes committed by Lucas were of the most gruesome and despicable nature, involving dismemberment, mutilation, rape, torture, pedophilia and even cannibalism! What did Lucas know that Bush didn't want people to hear? [38]

While conducting his presidential campaign in California in 1999, Bush was confronted by reporters about these incidents. He routinely refused to answer questions and would constantly change the subject! In the infamously beguiling words of his pro-Nazi, industrialist grandfather...

> *"There's three things to remember:*
> *Claim Everything, Explain Nothing, Deny Everything"*
> — (Senator Prescott Bush, Skull and Bones, 1966). [39]

Therefore, in keeping with the tradition of his war-mongering father and quisling Nazi grandfather, he blatantly "got away with murder"! [40]

1984: Brownsville, Texas 17 Murdered!

1994-2000: Bush as Governor of Texas 131 Executed!

2000: Presidential Election: Yes, he got away with murder, (Just Ask Al Gore!)

2001: "9-11" 3000 Killed!

2003: His Illegal and Unjustifiable Occupation of Iraq Up to 1 Million People Dead!

In the year 1555, Nostradamus predicted that in 1999, a "great ruler of terror" would arise and once again claim "King of the Mongols." Genghis Khan, previous holder of this title from the thirteenth century, had earned this designation from being the most prolific barbarian in history. He would routinely have his men rape, pillage, burn, and kill. His most notorious conquest in history was the battle of Baghdad in 1258, reportedly killing 500,000–1,000,000 people. During this era, Baghdad had one of the most extensive library systems in the world, which was destroyed, causing irreparable harm to academia, as well as to the advancement of civilization! How history repeats itself. In 1999, Bush was campaigning for office, therefore starting the election process, which inevitably resulted in the occupation of Iraq, where coincidentally, 500,000–1,000,000 people became victims of death from the unbridled tyranny of "Bush the Barbarian," the new "Ruling King of the Mongols." [41]

Note: It should be acknowledged that Nostradamus really "hit the nail on the head" with this one!

The supercilious and ethnocentric Bush family legacy serves as a prime example of the "robber baron" hiding behind Christian based "eschatological dispensations" * in order to justify their anti-environmental, war-mongering, and jingoistic protocol, which has driven our modern Western civilization for centuries! [42]

In 2007, Oprah Winfrey, was being sued by one of former President Bush's favorite organizations; The Texas Cattlemen's Association, for 12 million dollars, when she announced on her TV show, that she would *never eat another burger!*" However, based on the preponderance of evidence, the Judge ruled in favor of Oprah! "You can't sue someone for telling the truth!" (See chapter 12: "Mad Cows,Englishmen and Howard Lyman") [43]

Petroleum (petrochemical manufacturing, oil and gas exploration and production), along with beef (cattle ranching and slaughterhouse production), are two of the main industries which make up the economic infrastructure of Texas. There is no doubt that working in the oil industry is a dirty and hazardous occupation in itself. [**44**]

However, statistically speaking, slaughterhouse workers have the highest rate of on the job injuries, and a greater job turnover, than any other occupation! (Refer to chapters "The Malicious Mechanics of Modern Meat" and "PETA.") [45]

Green state vs. brown state: In 2007, while California made history by committing to a 25% reduction in global warming emissions by 2020, Texas, on the other hand, has built an additional 20 coal-fired power plants since that time. Statistically speaking, Texans already consume 60% more energy per capita than the average American. [46]

Imagine Texas one hundred years into the future, solar and wind farms replace dirty toxic oil and coal production. Clean, fresh, sustainable agriculture abounds, where only noxious, polluting, and cruel cattle ranching existed before. From "greedy oil barons," "smoke stack industrialists," and "coldhearted cattle tycoons" to sustainable and compassionate developers of energy and food. The infamous image of the brazen cowboy will fade into the history books like the extinction of the buffalo. Cowboys, cattle tycoons, oil barons, and dirty industrialists will become obsolete, nonpolitically correct figures, representing a bygone era of "conspicuous consumption." With the obsolescence of fossil fuels and coal, along with the sociology of modern sustainable food practices, then we will someday witness the generation of "the broken cowboy." But for now, if anyone asks you for directions on how to get to the "Ludicrous Lone Star State," just tell 'em to "go east till you smell it, and then go south till you step in it!"

It has been customary throughout history that soldiers have indulged in blood rare beef before battle. The American GIs during WWII were fed more than 130 grams of animal protein a day. This was equivalent to two and one-half times the average civilian quota. The impetus for this conclusion was contingent upon the popular belief that beef eaters are superior warriors and ascendant in terms of battle and survival in life-or-death situations. Russell Baker, social satirist of the 1920s, was attributed with the saying "beef madness," describing the high animal intake of the American soldier during WWI. The army's goal was to "beef up" every soldier into a "mighty, carnivorous killing machine"! [47]

> "As Long as there are Slaughter Houses, there will be Battle Fields"
> — (Leo Tolstoy)

Joseph Mitchell, staff writer for the *New Yorker Magazine* from 1938 until his death in 1996, describes an "era of excess" in New York City during the early part of the twentieth century in his classic book, "Up in the Old Hotel." He exemplifies the mind-set of this era by elaborating about a "gluttonous social event" called "the beef steak," where hamburgers were served as appetizers before the steaks. The average person at one of these voracious events consumed eight pounds of beef! Wow, eight pounds, that's like eating a baby! Just try to pound down eight pounds of the irradiated, hormone-adulterated, antibiotic tainted, artificially dyed, and pesticide- imbued beef of today! Can you spell "C-O-L-O-N-O-S-C-O-P-Y"? In the days when cars only averaged 4–6 MPG, this truly was an age of "conspicuous consumption"! [48]

By studying nature, it becomes obvious that herbivores are passive, meditative, and complacent creatures. As a rule, vegetarians (especially vegans), never commit any violent acts or crimes such as rape or mayhem. So, it's not surprising that a common phenomenon described by women, while being associated with the indignation of rape, was to "being treated like a piece of meat." After such an ordeal, a woman would tend to feel "emotionally butchered." [49]

Furthermore, this author is personally not aware of any (vegetarian), Buddhist- or Hindu-based organizations undergoing criminal and/or civil prosecution for its member's sexual improprieties. How could the Catholic Church, the most powerful, world-renowned religious institution in history, be plagued by such rampant scandal? The Roman Catholic Church has already paid out over 2 billion dollars in damages to its victims in the United States alone!

What type of sick and convoluted message does this send to our children, and say about our society? When sitting in a confessional, maybe the confession should start coming from the other side? This type of travesty undermines thousands of years of legitimate spiritual enrichment, leaving people ambivalent or jaded toward seeking true "empyrean guidance" (finding your higher power through divine inspiration). Therefore, it is this author's contention that the common denominator in all these "carnal indiscretions" is "meat," or what should be morally referred to as "devil's food"! [50]

In 2019 vegan celebrities, Woody Harrelson, Paul McCartney, Joaquin Phoenix and Evanna Lynch, along with plant-based physician, Dr. Neal Barnard), met with the Pope, in order to convince him to go vegan. [51]

According to the Bible, before there was "sin" in the Garden of Eden, Adam and Eve obeyed God's command to eat only foods produced by the earth. *"God said, 'I give you every seed- bearing plant on the face of the whole earth and every tree that has fruit with seed in it. They will be yours for food'"* (Gen. 1:29 NIV). Subsequently, it wasn't until after the "The Fall" that man began to eat meat.

The lesson here is ... "those whose dogma justifies indulging in the flesh suffers from a greater propensity of sinning in the flesh!" [52]

The "strutheous" [53] nature of our modern Western cattle culture is further exacerbated from generations of tradition through "mainstream theological justifications" which are in denial of true compassion, by relegating or eschewing the moral indignation of animal slaughter. People use their religions to conveniently justify their inhumane habits of animal consumption. Other than the people who may defame or corrupt it, there is nothing intrinsically wrong with any faith as long as it can be tolerant enough to incorporate a vegan mindset into its dogma! The definition of inhumane is also to be inhuman (Not worthy of being a human being!)

> *"When we do not intervene in long patterns of abuse, we tolerate and support that abuse – and our silence speaks our doctrine, doesn't it?"*
>
> — (The Reverend Tom Woodward)

The (vegetarian), Hindu-based, Vedic culture, is the world's oldest civilized culture stemming back over six thousand years. In their ancient language known as Sanskrit, the word for war also means desire for cattle! [54] During this era (meat-eating), Nomadic tribes would overgraze their herds for wealth and power and would therefore, infringe upon Vedic territories. Translated into English from ancient Vedic text, these "Aryan invaders" were referred to as "barbarians." The core meaning for the word barbarian stems from the phrase "to invade." This was the early civilized vegetarians' description of these barbarous meat eaters: "social scavengers," people of alien descent and usually believed to be inferior, crude, uncivilized, savage, uncouth individuals, lacking in refinement, erudite skills, artistic or literary culture. [55]

> "A nomadic, hoard of invaders cannot from any stretch of the imagination produce the kind of sublime wisdom, pure and pristine spiritual experiences of the highest order, a universal philosophy of religious tolerance and harmony for the entire mankind, that one finds in the Vedic literature."
> — (Lloyd Ridgeon) [56]

The heinous acts of marauding, pillaging, and raping are virtues which could only be born from nomadic flesh-eating cultures. The European settlers also expanded their herds into the new land, and just like their barbarian ancestors, neglected the human rights of millions of indigenous people! In today's society, three millennia later, the word beef still carries the same violent metaphorical connotation.[57]

The outcome of one's pugnacious demeanor, due to the engagement of a physical altercation, might be referred to as a "beef" or "getting into a beef." Being arrested and charged with a criminal offense is called "catching a beef." If inquiring into the impetus toward confrontation or cantankerous behavior, the question might be raised "What's the beef?"

Meat has also taken on a meaning of importance and substance. The essence of something is commonly referred to as "the meat of the matter." To improve or strengthen is to "beef up." Plant material often has an inverse word association. To "vegetate" or to be a "couch potato" generally gives an insipid, negative, unexciting or dull connotation. A "vegetable" describes the condition of a brain-dead individual.

Meat and gender rationalizations have also become apparent. For thousands of years, beef has dictated social order and gender hierarchy. Red meat and male dominance in a privileged hierarchy has been one of the longest-lived socio-cultural ethics which has survived throughout history. Historically, men have become so dependent on red meat that the absence of its provision by a spouse, due to the substitution of a vegetable product, has been a precursor toward violence against the woman.

Historically, men have been associated with war and red meat, while women represent plant material or more delicate types of flesh, denoting a more civilized demeanor. Men have been referred to by women as "hunks" or "beef cakes" while women are associated with such connotations as "chicks," "hot tomatoes," "peaches," or "wall flowers." [58]

Men have always dominated meat-eating cultures, while vegetarian societies have given more importance and egalitarianism to women. Anthropologist Peggy Sanday has concluded from a study of over one hundred non-technological cultures that animal-based economies are male dominated, while agriculture-based economies were greatly attributed to women. [59]

Accounting for the vast differences in composition and beliefs that separate diverse beef cultures, the greatest extent of them share an intrinsic relationship. Hunting and cultivating cultures have very distinct characteristics. Each have ascribed to a different paradigm of tradition and ethics, with diverse philosophical and religious dogma. Death, slaughter, and profligate waste manifest beef-eating cultures, while growth, conservation, and regeneration constitute vegetarian societies. Each culture has a distinct quality in their style of consuming the planet and of providing sustenance to their populations. To harvest or cultivate requires tending, growing, nurturing, and developing as compared to killing and slaughtering by quarry or sequestration. [60]

Therefore, the food we eat also plays a major role in the structuring of our moral values. The consumption of meat subconsciously modifies our perception of life and therefore impugns our moral fiber! After practicing veganism, we inevitably experience a metamorphosis which changes the way we think of ourselves and look at the world around us. It brings us to the realization of respect, true compassion, and indelible love for all living creatures. "Does something have to die in order for you to live?" Would you kill for food?

So in making this noble, conscientious, and compassionate decision to no longer participate in this bloody carnage, we not only feel better about ourselves, but we also benefit physically, while at the same time dramatically reducing our carbon foot print!

A typical Mid-western cliché thrown around for many years is that *"People who don't have the heart to kill their own food are weak and pathetic."* While we may not have the heart to slaughter animals, it's not due to our weakness but rather our strength. It is the strength of our spirit to move beyond carnal lust and therefore recognize love and compassion for all living creatures, which would morally prohibit us from such inhumane acts!

Our society today is one that has been dictated and predicated by "death and slaughter," and it is this mentality which has become bereft to the true nature of life itself. It is this dissolute blindness which could eventually lead to the end of the world as we know it! For man to survive into the twenty-second century, he must shift away from a mentality of "death and slaughter" to one of "growth and regeneration!"

Author Jeremy Rifkin writes ...

> *"The elimination of beef from the human diet signals an anthropological turning point in the history of human consciousness. By moving beyond the beef culture we forge a covenant for humanity, one based on protecting the health of the biosphere, providing sustenance for our fellow human beings, and caring for the welfare of the other creatures with whom we share the earth."* [61]

Plant products are environmentally safe, "karmic-ly-correct" embellishments bestowed to us by Mother Nature, sustenance we can ingest guilt free from the suffering incurred by living creatures, a spiritual food akin to manna. The anthropogenic, psychological/spiritual, and alimentary advantages of plant food are so conclusive that one could hardly understand how the modern industrial livestock complex could remain such an inexpugnable force? Plant cultures achieve a cosmic force with nature, one of inherent symbiosis between man and biomass (representing a more compassionate, sustainable, and altruistic lifestyle). Beef-eating cultures represent heartless evil, merciless death, brutal aggression, fascism, profligate destruction, narcissism, greed, hedonism, jingoism, male chauvinism, racism, demagoguery, militarism, industrialization, and pollution! [62]

The West has much to learn from the East in terms of diet and philosophy. The *American Heritage Dictionary* describes the word Ahimsa as "a

Jain Buddhist and Hindu doctrine of nonviolence expressing belief in the sacredness of all living creatures." [63]

Even Christian-based organizations understand and recognize the universal significance of this ancient Hindu/Buddhist doctrine:

> "Ahimsa is a Sanskrit word that basically means 'nonviolence in thought, word or action.' However it is much more. Because it is the positive counter force to Himsa (violence), it is dynamic proactive compassion towards all life. Since this precept can be found in the origins of faiths throughout the world, it transcends all religions. For human beings, Ahimsa is the first manifestation of an evolved conscience"
>
> — (Viatoris Ministries). [64]

Isn't this the reason why there are no vegetarian factions of Buddhist or Hindu terrorist organizations? Therefore, there should be no coincidence that all dangerous and violent factions of religious zealots, who kill in the name of God, all share one thing in common besides their propensity for mayhem, and that is to eat meat and morally justify the slaughter of animals!

> "Only when we have become nonviolent towards all life will we have learned to live well with others"
>
> — (Cesar Chavez)

Maybe if everyone who ate meat were required to participate in this bloody sacrilegious carnage—the hideous, wretched, fetid, raunchy, decomposing flesh, imbued with ensanguine properties; the disgusting putridity of entrails; the abhorrent, noxious, malodorous, festering stench from the effluvium emissions of the slaughterhouse—then they might change the way they feel about brutally taking the life of another living being, that like humans, cries out in pain, suffers, feels fear and pleasure, practices parenting and should have every right to exist on this planet free from the abattoirs blade!

Eating meat should be as politically correct as smoking tobacco. Whether or not people want to admit it, the consumption of beef represents an ignorant, selfish, and cavalier attitude. At least the tobacco industry doesn't intentionally kill animals or cause any significant amounts of global warming, drought, pollution, or famine like the cattle industry. However, smoking still does share some wonderful attributes with eating meat, like heart disease, stroke, and cancer!

Virtues of Veganism

- Longevity/Health Benefits
- Increased IQ
- Genuine Love for Animals/Compassion
- Ecological Benefits
- Nonviolent World View
- Psychological/Spiritual Benefits

Charles Darwin created a popular movement of evolutionary thinkers with his theory on "survival of the fittest," that man must take command over nature in order to be the fittest in his struggle over the elements. Darwinism was one aspect of socio-cultural ethics that justified the visceral nature of animal carnage, but it was the adaptation of the paradigm, known as "Social Darwinism," which transformed this hypothesis into one of racial supremacy based on the belief that European meat eaters were at the top of a social hierarchy. [65]

The European settlers believed that in order to gain superiority over other races and cultures, which they believed to be savages, they would have to be the fittest, the strongest, and most powerful. The inegalitarian, Eurocentric, evolutionary implications of Social Darwinism, during this era, are defined by *The American Heritage Dictionary* as *"the application of Darwinism to the study of human society, specifically a theory in sociology that individuals or groups achieve advantage over others as the result of genetic or biological superiority."* [66]

Therefore, the roots of prejudice in America directly stem from our European meat-eating culture! Since the writing of this literature, in 2019, in acknowledgement of the aforementioned atrocities committed against African and Native Americans, the holiday Columbus Day, was changed to Indigenous People's Day and is an official city and state holiday, in various regions of the United States. [67]

Analogy to Slavery about Slaughter:

Carol Adams, author of *The Sexual Politics of Meat*, clearly defines a visceral relationship between the historical European patriarchal role in regards to slavery and the dominance over Western civilization. She compares the practice of slavery to our meat-eating culture, stressing that as a species, just as we have evolved from the practice of slavery, humans must graduate to the "next stage of evolution" beyond the slaughter and suffering of animals! It's common sense that if your dogma prohibits you from abusing, killing, or eating animals (ahimsa), then in turn, how could you justify forcing other humans into slavery? Human nature dictates that people don't easily submit to slavery. They have to be physically and psychologically broken down— spirits broken, family ties lost—their servility must be nurtured through intimidation, brute force, and even torture.

> "But a growing number of people began to understand that because the abuses of slavery were upheld by custom, laws, and religion, the buying and selling of other human beings reflected a social standard as well as an individual choice. And they understood that as members of society which validated and perpetrated its cruelties, they showed moral responsibilities for its offences. Only when this happened did abolition become a possibility. And it is only when those who understand that killing other beings in order to satisfy an appetite for their flesh, is much more than a personal choice, that human carnivorism will become an anomaly. Only when the cruelty and immorality of breeding animals for the slaughterhouse is recognized as an evil sanctioned by society and upheld by its laws, will vegetarianism reach the next stage of evolution in western civilization."

– (Carol Adams) [68]

SLAVERY IN VEGETARIAN CULTURES:

While Buddhist and Hindu cultures do not go without entertaining certain forms of slavery, the severity of which should be mitigated in relationship to other meat-eating societies, which traditionally had practiced much more harsh and oppressive forms of slavery. Rather than to represent what most people would refer to as brazen slave labor, Hindu/Buddhist cultures practiced what is traditionally referred to as "obsequious servitude" (having a loving and grateful servant).

So, in essence, doesn't factory animal farming represent the worst form of slavery in itself? This is a cold and heartless environment in which vulnerable and defenseless animals' cries for help fall short of human empathy or understanding. These helpless creatures are slaves to a system of "coldhearted evil" with industrial "Soylent Green" imperatives! [69]

In conjunction with this racial hypothesis of slavery were the utilitarian justifications of anthropocentric Judeo-Christian theological ideals, which stated that since God created man in his own image, then the universe must revolve around man; thus, man has inherited the power over creation, and therefore, the animals and natural resources were placed on earth solely for his needs, absolving man of any guilt or conscience over the moral indignation from the slaughter of animals.

> "And God said, let us make man in our own image, after our likeness: and let them have dominion over the fish of the sea, and over the foul of the air, and over the cattle, and over all the earth and over every creeping thing that creepeth upon the earth." [70]

It was convenient for the early settlers to pattern their moral values around this first page in the Bible; however, they may have overlooked one other important Biblical proverb: "The meek shall inherit the earth." [71]

According to the Bible, in the story of Noah's Ark, even when man became too wicked for God's tolerance, animals were always sacred! [72] All animals are innocent in the eyes of God or nature. It's only humans that breed them, train them, or subjugate them in order to become live vehicles of exploitation!

> *"We know we cannot be kind to animals until we stop exploiting them; exploiting animals in the name of science; exploiting animals in the name of sport; exploiting animals in the name of fashion, and yes exploiting animals in the name of food."*
>
> — (Cesar Chavez)

Erasmus Darwin—renowned eighteenth-century British writer, physician, scientist, and social reformer—refers to nature as "one great slaughterhouse." Besides the obvious inference, this statement also describes the worldwide decimation and enslavement of indigenous people and the mass expropriation of land from Native Americans, all in the name of Christian reform and Social Darwinism, which was predominant during this era. [73]

Many Native Americans, and/or African Americans, who practice Christianity today may be offended by this literature. Dogmatically speaking, to some, it could be interpreted as blasphemous. And if this is the case, may this author humbly apologize. However, what these sincerely misguided and gullible individuals do not realize is that many of their ancestors (black and Native American), were forced their religion and surnames from the "Eurocentric" [74] settlers who decimated, subjugated or enslaved their forbearance. For example, the names Perez and Washington, which are common among Hispanic and Black people, are both derived from European ancestry.

In other words, atrocities, in regard to human rights were committed in the name of God, against other races and cultures that were not white or Christian. White Christians were the predominant factor in the creation of prejudice in this country. Hate and animosity against Blacks and Native Americans is an age-old Christian ethic! There was nothing wrong with Christ, it was just many of his followers who have historically exhibited and justified their deplorable behavior in his name!

Many political regimes throughout history have been hidden behind a shroud of religion in order to justify genocidal acts against humanity. "Good People do Good things. Bad people do Bad things. But when Good people do Bad things, it's because of Religion" — (Dr. Bernard Haisch). The real underlying agenda has been predominantly for greed and power, while the peasants or working class are led to believe that it's a holy war! [75]

Atrocities against humanity, committed in the name of God, historically, are only born from meat-eating cultures. This anthropocentric justification to go out and conquer in the name of God has been the basis and foundation for the legacy of our jingoistic behavior in America for hundreds of years!

"For what were all those celebrated conquerors but the great butchers of mankind."
— (John Locke) [76]

Antisocial Factors of Meat-Eating Cultures

- Anthropocentrism
- Speciesism
- Demagoguism
- Hedonism
- Jingoism
- Male Chauvinism
- Militarism
- Racism

- Terrorism
- Sexism
- Narcissism
- Imperialism
- Negative Consumerism

HISTORIC ANTISOCIAL FACTORS ASSOCIATED WITH MEAT-EATING CULTURES

- Barbarianism
- War Mongering
- Conquering
- Killing
- Slaughtering
- Raping
- Plundering
- Subjugating
- Enslaving

"Veni Vidi Vici" (They Came, They Saw, They Conquered!).

Now if white supremacy is no longer politically correct, then why do so many people of color as well as Native Americans neglect the sacred roots and heritage of their ancestors? How time makes people forget. The European settlers came to the new world with the mind-set that if indigenous peoples were not white or Christian, then their lives had no value. The Native Americans, as well as the African people, were an expendable commodity by the European settlers, as they were considered "savages" and like "animals" were not worthy of human rights. Therefore, up until the mid-1800s, it was still an acceptable practice to keep slaves and slaughter Indians!

One of the most notorious "Indian killers" of American history was President Andrew Jackson. During his reign in office, not only did he believe in and practice slavery, but he also publicly and openly loathed and degraded Native American citizens, exterminated, or exiled them from their ancestral lands, and showed absolutely no respect for their heritage or human rights!

Imagine, in the middle of a deadly winter, a heartless and unconstitutional Jacksonian federal mandate seizes an already subjugated and provoked Cherokee Nation, illegally forcing them from their rich and bountiful sacred lands, only to be relocated in worthless, barren, and desolate territories west of the Mississippi. More than four thousand Native Americans died or became victims of this tyranny. Even the "civilized" Cherokees (who had completely assimilated to the "white man's" standards), were also forced to leave Georgia based solely on the color of their skin! Hence, from this tragic journey, the Cherokee people left behind what has been infamously referred to as the "Trail of Tears."

Avarice was a major factor by the white settlers in their motivation to eliminate Native American populations. The settlers felt that the Indians were impeding or standing in the way of progress. With the utilization of the transcontinental railroad, along with the invention of the cotton gin and other modern technologies, there was a growing demand for agriculture and ranch land, as well as various forms of commerce.

In 1830, the state of Georgia, with strong support from President Andrew Jackson, effectively confiscated all Cherokee land. Thereafter, the plight of the Cherokee people was legally defended all the way to the highest courts in the land. The final Supreme Court decision ruled in favor of the Cherokee Nation. A mandate was ordered by Supreme Court Justice John Marshall and sent to President Jackson, which ruled that federal treaties with Native Americans were legally binding and that by federal law, Georgia had no legal right to confiscate Cherokee land.

However, in a written response to this Supreme Court ruling, President Andrew Jackson displayed gross malfeasance against the Cherokee people, stating that Supreme Court Justice "John Marshall has made his decision; now let him enforce it!" This particular incident is just one of many throughout American history, too numerous to count, which represents a legacy of Native American injustice. According to author Robert Lindneux's rendition of the Trail of Tears...

> "These types of policies raped the Native American of his soul and land, in an age when supposed civilized behavior reigns supreme. By stealing property in order to redistribute the land to Euro-Americans, the Government contradicted sacred property rights." [77]

Ask any Native American, and they would tell you that our history books are filled with lies to make our white Christian forefathers look like heroes, instead of the plundering barbarians that they actually were! The irony of our culture is that to a "real American" (a Native American), having Jackson's picture on a twenty-dollar bill is equivalent to putting Adolf Hitler's picture on Israeli currency! How would you feel?

Committing genocide on millions of buffalo, in itself, is an egregious anthropogenic travesty, one which will live in infamy in the hearts and minds of all Native Americans as well as those who have compassion for living creatures and wish to preserve species on this planet. As is evident in history, this "Bovinical" decimation was a staggering blow to all Native Americans from which they have never recovered! Obviously, only a meat-eating culture could perpetrate such barbaric and uncompassionate behavior!

In a new book entitled *Ecotopia*, the author has created a modern and more politically correct version of the Ten Commandments called "The Earth's Ten Commandments." In this literature, the third commandment states, "Thou shalt not hold thyself above other living things nor drive them to extinction." [78]

KILLING THE BUFFALO CRIPPLED THE INDIANS IN TWO WAYS:

1. It crippled them spiritually: These truly cowardly and sacrilegious acts, destroyed their belief in the "Great Provider." They felt shame and guilt over the disappearance of this once- prolific and mighty ungulate. As if they had somehow desecrated the "Great Spirit," and this broke their spirit, thus succumbing to the "white man" and the ways of the "white man."

2. It crippled them economically: The buffalo provided for all of their utilitarian needs: food, clothing, and shelter. Once the buffalo were gone, they were almost completely dependent on the "white man," especially after already being slaughtered and subjugated for hundreds of years! [79]

And as far as aboriginal Americans are concerned, for the terrible things our forefathers did, we have given casinos to some elite Native American tribes who, statistically speaking, may not be sharing this newfound wealth equally with other tribes who may have been equally oppressed over the centuries. How does this make up for the many hundreds of years of malfeasance against all indigenous Americans?

To give more credence to this theory, consider that up until the late 1960s, Asian, Hispanic, Native, and African Americans were still considered second-class citizens, living under an antiquated caste system, not befitting of a true democracy in the twentieth century "where all men are created equal!" Even for more than one hundred years after Lincoln freed the slaves, this factor was still exploited until Martin Luther King finally delivered his powerful civil rights diatribe! Now more than forty years later, we finally had a black president. This would have all those old Southern white "crackers" turning over in their graves!

During the civil rights era, it was evident that blacks and Hispanics were usually never given leading character roles in TV or cinema, and any exposure was usually characterized or portrayed in a stereotypical, derogatory, or perfunctory manner. But even an anomaly, like the great black actress Dorothy Dandridge, who was a top-paid movie star in her day, got a sober awakening to the reality of her era when she playfully dipped her toe in a Las Vegas hotel swimming pool after being warned by management about hotel policy, and to her chagrin, the pool was mandatorily drained and scrubbed down. Ironically, by black laborers!

Remember, this was Las Vegas, not the Deep South, yet after one hundred years of supposed freedom from the oppression of slavery, the same white, racial Christian-based mentality still existed! It was hard to forget the look of gaping consternation on her face when she realized the reality of her situation. It didn't matter who she was, she was black, and that was all that mattered! [80]

The genocide of Indians, buffalo, and the enslavement of Africans was acceptable behavior only 150 years ago! What a bunch of rowdy hooligans our forefathers were. And we have the gall to make fun of the Australians? How can we be proud of our heritage? And out of all this racism and bad behavior was born a legacy of right wing, eschatological, Southern white Christian GOP protocol!

RIGHT WING, CHRISTIAN-BASED, WHITE SUPREMACIST THEORY:
The basis for the Ku Klux Klan today represents the same antediluvian Southern white Baptist Christian mentality of Confederate slave owners carried over from the Civil War. (Refer to chapter "Peace and Prosperity.") [81]

Another offshoot of these same eschatological* Christian-based, neo-Nazi political ideologies is the Lyndon LaRouche Organization. But they take it one step further; they were actually promoting the end of the world themselves by endorsing nuclear power, and in doing so, don't they create their own self-fulfilling prophecy of the apocalypse? Lyndon LaRouche was running for president during the 2008 presidential election. His organization was heavily supported by the nuclear power industry. These "Nuclear Nazi's" were literally "purveyors of hate and Armageddon!" (Refer to chapter "The Insidious Nature of Nuclear Power.") [82]

> *"Armageddon is coming. They can sign all the peace treaties they want [in the Mideast].They don't do any good. There are dark days coming. My Lord, I'm happy about it. He [Jesus] is coming again … I don't care who it troubles. It thrills my soul."*
>
> — (Jimmy Swaggart) [83]

At midpoint into the cold war crisis, sensing people's fears, author Hal Lindsey sparked Christian-based apocalyptic speculation with his book *The Late Great Planet Earth*, which sold over 10 million copies. [84] In 1983, Billy Graham, who had great credibility during this era, spurred new controversy regarding millennial expectations in his book *Approaching Hoof Beats: The Four Horsemen of the Apocalypse*, in which he extrapolates about negative aspects of politics and society, stating that "Jesus, the Man on the White Horse," will come when man has sunk to his lowest level. [85] In addition, Ronald Reagan, in 1983, also bolstered this "mainstream eschatological mind-set" by citing scriptural authority in his justification of labeling the Soviet Union as an "evil empire"! [86]

In 1986, Grace Halsell, came out with a dynamic book entitled *Prophecy and Politics: Militant Evangelists on the Road to Nuclear War*, which outlines how so many evangelists— such as Pat Robertson, Jerry Fallwell, Billy Graham, and Hal Lindsey, to name a few—had been capitalizing on this end of the world mania, all insinuating in their literature that nuclear holocaust will represent the final scenario in the "end of times"! [87]

> *"Theocracy plus nuclear weaponry is a terrifying threat."*
>
> — (Christopher Hitchens)

*Eschatological Dispensations: The end of the world theory has been av-
ariciously capitalized on by U.S. politicians, and American evangelists over
the last one hundred years, through Christian-based "dispensations." Many
(GOP) politicians have historically used dispensationalism to justify their
jingoistic protocol and negligent environmental policies. (Refer Trump,
Reagan and Bush policies in chapter "Peace and Prosperity").

> "Apocalyptic fears and millennial expectation play an important role
> in three sectors of right-wing populism in which demonization, scape-
> goating, and conspiracism flourish: the Christian Right; the populist
> right, including survivalist, Patriot, and armed militia movements;
> and the far right, especially the neo-Nazi version of Christian Iden-
> tity theology. In the 1970s and 1980s far right Christian Identity and
> Constitutionalist groups interacted with apocalyptic survivalists to
> spawn a number of militant quasi-underground formations, including
> some that called themselves patriots or militias. During the height of
> the rural farm crisis in the early 1980s, one of these groups, the Posse
> Comitatus—a loosely-knit armed network that spread conspiracism,
> white supremacy, and anti-Semitism throughout the farm belt—cap-
> tured a small but significant number of sympathizers among farmers
> and ranchers. Other groups, such as Aryan Nations and the Lyndon
> LaRouche group were also active, and soon a loose network was con-
> structed linking tax protesters to groups as far to the right as various
> Ku Klux Klan splinter groups and neo-Nazi organizations."
>
> — (Publiceye.org). [88]

Could it be logically remonstrated that eschatological dispensations represent the most negative of all human social traits: "misanthropy?" Like as
in the story of Noah's Ark, the rapture represents "God's hatred of mankind"
and his inevitable destruction in order to cleanse the world of wicked and
evil and therefore reestablish new world order! However, the irony here for
those Christians who subscribe to the literal interpretation of the Bible is
that in the story of Noah's Ark, God clearly gives his word to Noah that he
will never again destroy all living things on earth: "And the Lord said in his
heart, I will not again curse the ground anymore for mans sake; for the imag-
ination of man's heart is evil from his youth; neither will I smite any more
everything living, as I have done." [89]

Doesn't this quote directly debunk all subsequent eschatological theory? Eschatology should be referred to as E-scatology (Scatology: the study of fecal matter). Because when it all comes right down to it, eSCHATology is all a bunch of "Bull SCHAT"!

How can a self-destructive and negative belief (the hatred and destruction of mankind by God), be used in a constructive manner to help better serve man in the future? As history has clearly demonstrated, this eschatological propaganda has been used as a tool by politicians to justify a legacy of racial supremacy, military defense spending, and anti-environmental practices! This truly is "the psychology of the cattle culture"!

IN THE TEACHINGS OF KRISHNA (HINDU-BASED PHILOSOPHY), THEY SPEAK OF THE FOLLOWING:

<u>Four Major Sins of Man</u>:

1. Eating Meat
2. Illicit Sex
3. Illicit Drugs
4. Gambling

According to their traditional doctrine, the mere act of eating meat puts your soul through negative transgressions, thus lowering your vibrations and making yourself more vulnerable to the other three vices. When you analyze all the carnage that each human being is responsible for in their lifetime, you begin to understand how "karma" plays an important role in the advancement of our spiritual self. They believe that ingesting flesh makes a person less demure, esoteric, or spiritual, and therefore, more aggressive, worldly, materialistic and narcissistic.

These are all self-indulgent traits characterized by our meat-eating culture. [90]

It can also be scientifically proven, beyond a reasonable doubt, that creatures with two eyes, two ears, and one nose, who bleed red blood, suffer when slaughtered. There is no substantial evidence that plant material experiences any similar trauma. In Hindu/Buddhist doctrine, all plant material is believed to share one Great Spirit. But as we move up the evolutionary ladder, it is believed that mammals have only one soul. This theory is not so esoteric that it doesn't make logical sense!

And the Bible concurs: In literature sponsored by all-creatures.org and presented by Viatoris Ministries (a Christian publication and resource ministry), they speak of "God's covenant with animals," which they believe has been conveniently overlooked by traditional Christianity for thousands of years. The Bible proclaims that both animals and humans are "Nefesh Chaya" (living souls). In the book of Revelations, it repeatedly refers to both animals and humans occupying the throne of God. Therefore, it should be clear to any theologian who translates the Bible that animals were endowed by their creator with a soul! [91]

In regard to this biblical translation is a recent work by author J. R. Hyland, "What the Bible Really Says." It compares an ambiguous dichotomy of elements based on conventional and academic interpretation of scripture to a logical conclusion of "what the Bible really says." [92]

WAS CHRIST A VEGAN?

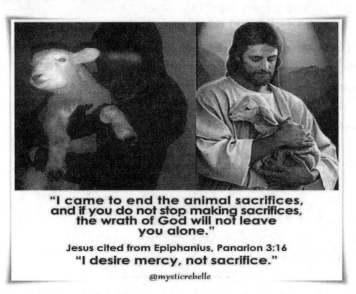

"I came to end the animal sacrifices, and if you do not stop making sacrifices, the wrath of God will not leave you alone."

Jesus cited from Epiphanius, Panarion 3:16

"I desire mercy, not sacrifice."

@mysticrebelle

Today in mainstream Christianity, there is a growing movement toward vegetarianism. However, there are still many stodgy or old traditional Christians who believe these people to be eccentric, or even troublemakers, who are trying to introduce a "New Age" lifestyle to traditional Christian ethics. But the Dead Sea Scrolls prove that vegetarianism is not new at all. This noble virtue was practiced by Christ and the earliest of Christians!

> *"But when those who leave traditional Christianity find a group who incorporates the principle of reverence for all God's creatures, they remain our companions in the effort to reestablish the divine order of love and compassion on earth. They to, are praying for a world that will reflect God's goodness. And they too, are trying to live the kind of lives that reflect the heavenly kingdom, in which no being is abused or unloved. Perhaps we can learn from each other how best to reach those among us who do not yet understand the ungodliness of their treatment of non-human beings."*
>
> — (Viatoris Ministries) (93)

"In a recent discovery (as in terms of Biblical chronological order), in the 1950s, of the Dead Sea Scrolls, it gives strong evidence that Christ was actually a vegan who taught and lived by the principle of nonviolence. It also makes reference about some of the original Christians called the Essenes, who lived by these same principles."

"The passages in the Bible have been completely corrupted, being written by 70 meat eating men, but NO women and translated 5 times over 2 thousand years through several meat eating kingdoms! The Dead Sea Scrolls were only translated once from Hebrew when referring to Christ and identifies him as one of the original Christians called the Essenes! Josephus, the most credible 1st century historian described the Essenes in one of his many volumes of history. "They [The Essenes] lived on vegetables that grew on their own accord, and wore such clothing as could be procured from plants and trees." —(Josephus, 1st century historian)"[94]

ANIMAL SACRIFICE:

Early man didn't eat meat every day, like as in today's modern Western culture. Meat was only used for special religious occasions or holidays. During this Biblical era, animals were still considered sacred and were therefore anointed by priests as sacrifice to the God's. However, in order to keep up with the growing demand of sacrificial meat, due to the patronization of temples by a burgeoning populous, priests became the official butchers of their community and their sacred shrines, slaughterhouses!

"In our own day, we have never been faced with the realities of sacrificial religion. To modern minds, phrases like the "holy of holies" and the "altar of the Lord" conjure up some sort of ancient, godly society in which priestly men served the Lord in His sanctuary. But the Temple was actually a slaughterhouse, and the linen vestments of the priests were stained red with the blood of their victims. And terms like the "altar of the Lord," obscure the fact that what is being talked about is the killing and butchering of helpless victims, in the name of God."

— (Viatoris Ministries) [95]

According to the scriptures, **Christ** was sentenced to death for his part in preventing the execution of tens of thousands of sacrificial lambs who would have been killed on Easter Sunday.

"Many who honor Jesus Christ celebrate his escape from the grave by condemning to death tens of thousands of gentile lambs. They will be brutally slaughtered so their corpses can be devoured on Easter Sunday." [96]

Jesus freed the animals in the temple which in reality was actually a butcher shop, thus declaring "I desire MERCY not SACRIFICE!" He subsequently was arrested for being an animal rights activist. [97]

All the great prophets of Israel (Amos, Isaiah, Hosea, and Jeremiah), had spoken out for the same compassionate reasons as Jesus, also condemning this practice of ritualistic animal slaughter; but like Christ, their cries were all in vain!

"When you offer me holocausts, I reject your oblations and refuse to look at your sacrifices of fattened cattle . . . but let justice flow like water and integrity like an unfailing stream.

— (Amos) [98]

"What are your endless sacrifices to me, says Yahweh. I am sick of holocausts of rams . . . the blood of bulls and goats revolts me . . . the smoke of them fills me with disgust . . . Your New Moons and your pilgrimages I hate with all my soul . . . your hands are covered with blood, wash, make yourselves clean."

— (Isaiah) [99]

"All living creatures may lie down without fear"
— (Hosea) : (2:18) (NEB) [100]

"Nothing in all creation is hidden from gods' right."
— (Hebrews) : (4:13) [101]

In the "Gospel of Thomas," Jesus makes "Profound Statements of Wisdom and Compassion" which are more reflective of Buddha, leading some theologians to question the authenticity of these statements. Another popular belief among the aforementioned authors, is that, like with many of the Israelites, Jesus was also "Specifically opposed" to sacrificial practices. Jesus was adamant over the issue of animal rights which entailed the issue of "Idolatry" (Whether one should eat meat sacrificed to idols). [102]

Clement of Alexandria (1st - 2nd century Christian theologian), had also made it evident that "Sacrifices were invented by men to be a pretext for eating flesh." He also wrote "It is far better to be happy than to have your bodies act as graveyards for animals." All the Apostles were vegan including Paul, who later converted. [103]

Historically, this debate had been quite significant, especially in light of powerful evidence which suggests that this is what led to Christ's demise! At what point in history did Judeo-Christian ethics move past the point of sacrificial empathy and sacred principle to one of utilitarian justification for the slaughter of animals? In the year 300 AD, Constantine officially allowed the consumption of flesh in his kingdom. Before this time Christians were supposed to be vegetarian. [104)

Compassionate author Jeremy Rifkin gives brilliant insight on this subject in his book *Beyond Beef* in the chapter appropriately entitled "Sacrifice to Slaughter."

> *"Cultures have further separated themselves from the animals they eat by shifting blame for their deaths, concealing the act of slaughter, misrepresenting the process of dismemberment, and disguising the identity of the animal during food preparation."*
>
> — (Jeremy Rifkin) [105]

Jesus went to his death refusing to use violence against his captors in order to extricate himself from this fatally violent act against him! In the Dead Sea Scrolls, it clearly states that Christ's last words to his disciples were ...

"Those who live by the sword shall die by the sword." [106]

Over the years, many authors and historians have also expostulated, their beliefs that Christ was an ethical vegan! The list includes Rynn Berry, (world renowned historian), authors, Charles Voclavic, Keith Akers, Carl Anders Skiver, Steven Davies, J. R. Hyland, Paramahansa Yogananda, and several others.

If your car breaks down, do you go to a preacher, a minister, a rabbi or a mechanic? A mechanic right? So, when one wants to learn a PROFESSIONAL non-bias opinion about the life of Christ, then they should go to a HISTORIAN, not a religious figure! Because, if you go to the three aforementioned religious figures, they may tell you 3 different stories!

So, when the Vegetarian Society hired renowned historian, Rynn Berry, Cornell graduate and ancient studies scholar, to research the life of Jesus, he concluded that Christ was a "Vegan Animal Rights Activist," and carried no weapons. Christ was sentenced to death for saving the lives of 10,000 sacrificial lambs who would have been brutally murdered for Easter Sunday.

Quotes taken from an interview with Rynn Berry in 2009:

Caryn Hartgrass: "... Was Jesus a vegetarian"?

Rynn Berry: "I mentioned that he was, and in my various books, I make a case for him being a vegetarian. After all it was difficult to not be a vegetarian in ancient Israel. The average person lived on a plant based diet out of economic necessity. Animal flesh was a luxury and Jesus certainly emphasizing simplicity of living would have had to be vegetarian, but also had ethical precepts very similar to those of Buddhism in which "Thou Shalt Not Kill" is really the first commandment to the Essenes of which he belonged. All the evidence suggests that he was a vegan!"

Caryn Hartgrass: "Right he was protesting with the slaughter in the temples."

Rynn Berry: "Right, that's probably what got him killed as he was protesting the animal sacrifice in the temples. You know at that time, people would go into the temple and purchase an animal and the priest would butcher it, then it would be roasted on the alter and shared among the worshipers. And so animal sacrifice was connected with flesh eating, soon as he did that he was brought up on charges and executed!" [107] & [108]

"Until he extends the circle of compassion to all living things, man will not himself find peace"

– (Albert Schweitzer, world renowned Christian philanthropist, humanitarian, and vegetarian). [109]

In the book *Disciples: How Jewish Christianity Shaped Jesus and Shattered the Church,* by Keith Akers, the author points out the absence of lamb on the table at Jesus's Passover celebration. He also insinuates, due to so many ambiguous translations of biblical text over the years, that there is no valid justification to assume that Jesus ate fish. Fish are such a universal symbol of abundance; therefore, current Biblical translation may only reflect hyperbole. [110]

Steven Davies, Author of *The Lost Gospel of Thomas* refers to Biblical passages taken from "The Gospel of Thomas." In saying 11, it states *"The dead do not live and the living will not die when you ate dead things (Animals) you made them alive. When you arrive into light (Achieve higher consciousness) what will you do?" "When you were one (Adam), you became two (Adam and Eve) when you become two, what will you do?"* (The abstinence of eating flesh) Adam and Eve only ate the food provided in the garden until the "Fall." Also, nowhere in the New Testament is Jesus depicted as eating flesh!

In saying 12, when the disciples ask Jesus, who will replace him as the head of the church, he replies "My brother James." "James the Just of course!" James is prolifically recognized by many credible ancient scholarly sources, as an ethical vegetarian! Remember, this division in the human species, ultimately led to what is referred to as the "Fall," where people started to eat flesh! Stated from the *Church History of Eusebius,* book 2, chapter 23: "James drank no wine or strong drinks, nor did he eat flesh!" — (Hegesppus) [111]

St. Basil was also quoted to have said, "The steam of meat meals darkens the spirit. One can hardly have virtue if one enjoys meat meals and feasts! In early paradise, no one sacrificed animals and no one ate meat!" Saying 111 also gives this foreboding warning to mankind! "Anyone living from the living (plants) will not die." This decree also reinforces saying 87: "Wretched is a body depending on a body (Corpse), and wretched is a soul depending on these two" Reinforcing this last statement is Saying 112:

> "Woe to the flesh dependent on the soul, woe to the soul dependent on the flesh!" [112]

St. Augustine also tells a similar story about Jacobus who lived on seeds, vegetables, and also refused meat and wine. [113]

Other passages of the Bible also support vegan ideals, such as in the story of Daniel. Daniel refused meat and the rich foods of the king, such as Cheese, and requested a simple diet of vegetables so that he might not defile his spirit [114]

Another second century work refers to the teachings of St. Peter: Homily XII states "The unnatural eating of flesh meats is as polluting as the heathen worship of devils, with its sacrifices and its impure feasts, through participation in it a man becomes a fellow eater with devils." [115]

St. Francis, makes this statement: "All things of creation are children of the father and thus brothers of man. God wants us to help animals, if they need help. Every creature in distress has the same right to be protected." [116] During this era, St Jerome, Catholic writer of the "Vulgate" Bible, also concurred with St. Francis's compassionate revelations. "The use of the flesh of animals was unknown up to the Deluge; but after the Deluge, men put between their teeth the sinews and stinking juices of flesh!" — (St. Jerome) [117]

In a book entitled *The Yoga of Jesus*" it is also explained that Christ led an ascetic vegetarian lifestyle and taught meditation. Information for this book was compiled from the Second Coming of Christ by the late Paramahansa Yogananda, founder of SRF (Self-Realization Fellowship Church). [118]

In addition, many other books also reinforce these theories. Some of these titles include *The Lost Beginnings of Creation and Christianity* by Charles Anders Shriver, *What to Serve a Goddess When She Comes to Dinner* by James R. Deal, *The Origin of Christianity* by Charles Vaclavic, and *Jesus Against The Scribal Elite* by Chris Keith. [119]

Many traditional meat-eating Christians may get offended when you question the literal interpretation of their scripture as opposed to a metaphorical approach. So then why is it when you read one of the most significant, if not the most prolific, Biblical decrees of all time, "Thou Shalt Not Kill," there is no disclaimer afterward in parentheses, which states "Unless you're having a barbecue"? Therefore, doesn't this give a hypocritical connotation toward any meat- eating Christian who subscribes to literal interpretation?

Isn't being a good Christian represent emulating Christ in his thoughts and good deeds? To be "Christ-like," acting in a manner which is befitting to Christ? Remember, "the will of God will not lead you where the grace of God cannot keep you." — (James Wallis). Therefore, how can a spiritually evolved human being, without conscience or remorse, brutally murder in cold blood one of God's precious creatures? So to answer your own dogmatically related query, "Should I eat meat?" Then consider, "if Jesus and the devil were standing in the road and a lamb crosses the road, which would kill the lamb?"

In theory, while religious doctrine may be sacrosanct, monolithic, organized religion is far from it! Organized religion needs a revenue stream to operate and therefore, just like any other business, is subject to corruption and scandal. This authoritarian view of religion displays a mind-set which not only compromises compassion for living creatures, but also creates an environment in which only black and white can exist; thus, under the precepts of scripture, priceless scientific knowledge may be grossly discounted. Traditional organized religion wants to teach people a generic form of faith. A one-size-fits-all mentality in order to justify the nefarious empirical designs by the powers that be over a naïve and credulous macro populous! If we truly are a civilized and compassionate society, then we will examine the facts and move beyond our superficial monolithic dogma and do what is right for our planet, our children, and our spiritual well-being!

So, in conclusion: "Don't let your dogma get run over by your karma!"

***

Obviously, in our aggressive meat-eating American culture, due to nuclear proliferation and modern warfare technology, we have become masters in the art of offence. And boy, are we ever offensive! So much so, that the legacy of our ostentatious, aggressive and jingoistic nature has internationally earned us the title of the "*Ugly American*." [120] However, legend of popular folk heroes suggests that the humble and self-deprecating, vegetarian Buddhist monks (shaolin) [**121**], have always been, and still remain masters in the art of defense (self-defense). So remember, no matter how big and tough you think you are by eating all that red meat, there's still always a vegetarian monk somewhere out there, half your size, who could "take you down to Chinatown!"

Another words, grasshopper, if unduly provoked, these innocuous-looking and peaceful Buddhists could theoretically transform into "boot-ists" and thereby proceed to kick your big fat "GluttonousMaximAss"!

And speaking of vegan heroes, Popeye is the world's most prolific vegan! Under this genre, his macho acts of heroism are motivated by the ingestion of spinach, yet the only one eating a hamburger is referred to as "Wimpy"! [122]

Now take for example our modern-day military complex. It is this same mentality that has built our Western empire, by war and subjugation. To be the strongest and most powerful, but like the Roman Empire, we are headed toward self-destruction.

The modern industrial cattle complex also shares a synonymous relationship with our military defense industry. Both are quite lucrative and represent death, slaughter, and profligate destruction; billions of dollars have been spent to develop and construct weapons of Armageddon; millions of U.S. citizens have earned their livelihood manufacturing these implements of genocide; yet the irony here is that due to the naive and recondite nature of public awareness, cattle could destroy our planet before these weapons of doom! However, the same man whose brilliant intellect, which had harnessed the power that could destroy the world, also possessed the knowledge and wisdom which could save it!

> *"Nothing will benefit health and increase the chance for survival on earth as much as the evolution to a vegetarian diet."*
>
> — (Albert Einstein) [**123**]

Notes:

[1]   The great Bull God Apis: A great bovine God worshiped in ancient Egyptian mythology.

[2]   Jeremy Rifkin, "Beyond Beef," (92: 166, 246)

[3]   Subreption: Any malicious or deceitful misrepresentation or concealment of facts; disinformation or lack of information; concealment of the truth (a powerful tool which governments, and the corporations who control them, effectively use, by propagandized distortion of media coverage or through nonexistent media coverage, about certain issues that we should all be aware of, but are not in the best interest of big business to tell us!).

[4]   Jeremy Rifkin: "Beyond Beef," (92:251–56).

[5]   Jack in the Box Defames Vegetarians: www.bayareaveg.org/forum/viewtopic. php?t=783.

[6]   United Nations report, "Livestock's Long Shadow-Environmental Issues and Options, 2006, by the Food and Agriculture Organization of the United Nations," is the most geospatially comprehensive, state-of-the-art report, which examines the global environmental impact of livestock production. www.fao.org/newsroom/en/ news/2006/1000448/index.html.

[7]   Jeremy Rifkin: "Beyond Beef," (92)

[8]   http://www.worldwatch.org.

[9]   www.earthshare.org.

[10]   www.earthsave.org.

[11]   earthshare.org

[12]   http://www.siwi.org

[13]   United Nations report, "Livestock's Long Shadow-Environmental Issues and Options, 2006, by the Food and Agriculture Organization of the United Nations," is the most geospatially comprehensive, state-of-the-art report, which examines the global environmental impact of livestock production. www.fao.org/newsroom/en/ news/2006/1000448/index.html

[14]   www.vegsoc.org

[15]   Story of a Cowboy, 1897, Emerson Hough

[16]   www.vegetarianvegan.org

[17]   In 1966, Charles Joseph Whitman, student at the University of Austin in Texas, while positioned inside a tower at the university, went on a shooting rampage, killing twelve people and wounding thirty-two others. http://74.6.117.48/search/srpcache?ei=UTF8 &p=University+of+Austin%2c+-Texas+sniper+attack&.

[18]   1963, The assassination of President John F. Kennedy in Dallas, Texas. http://en.wikipedia.org/wiki/JFKassasination.

[19]   http://guides.gottrouble.com/Glossery_of_Murder_Terms_Texas-r1204907-Texas.html

[20]   http://behavioralhealthcentral.com/index.php/20100226211173/Legislative-News/top-states- with-greatest-number-of-sex-offenders

[21]   Ethnocentrism: Characterized by, or based on a belief that one race or group of people is superior over others in terms of European culture.

[22]   http://www.anus.com/etc/texas.

[23]   http://www.livius.org/pi-pm/pilate02.html.

[24]   http://www.texecutions.com.

[25]   Texas Executions: "GW Bush Has Defined Himself, Unforgettably, as Shallow and Callous," by Anthony Lewis, 6/17/2000, New York Times.

[26]   http://en.wikipedia.org/wiki/Capital_punishment_in_the_United_States#History.

[27]   https://www.aljazeera.com/opinions/2019/3/20/why-did-bush-go-to-war-in-iraq

[28]   John Paul Stevens, retired Supreme Court Justice. http://www.cbsnews.com/ stories/2010/11/23/60minutes/main7082572.shtml.

[29]   http://movies.about.com/cs/texaschainsaw/a/texasintth.htm.

[30]   Speciesism: http//www.merriam-webster.com/dictionary/speciesism

[31]   Anthropocentrism: http//www.merriam-webster.com/dictionary/anthropocentrism

[32]   http://cats.about.com/cs/crueltyconnection/a/cruelty.htm.

[33]   http://www.mspca.org/assets/documents/annual-report-2009-2.pdf.

[34]   http://www.pet-abuse.com/pages/abuse_connection.php.

[35]   The Silence of the Lambs: http://cinema.theiapolis.com/movie-2rml/the-silence-of-thelambs/storyline.html.

[36]   http://www.zenzoneforum.com/threads/7019-George-Bush-and-the-1984-Brownsville-Mass-Murder

[37]   A CHARGE TO KEEP: My Journey to the White House. 1999, G. W. Bush, IISBNO-688- 17441-3 William Morris.

[38]   http://portland.indymedia.org/en/2005/01/308911.shtml

[39]   "Bush Bones Doctrine": http://www.nogw.com/dark_history.html.

[40]   http://www.nogw/images/bush-dynasty.gif.

[41]   The New Genghis Khan, http://www.nogw.com/dark_history.html.

[42]   http://watch-and recap.com/? s=84urq6u2q53n24tglid2u4u4u4-!.

[43]   https://www.britannica.com/story/a-brief-history-of-food-libel-laws

[44]   www.slideshare.net/mechportal/occupational-hazards.

[45]   http://www.sustainabletable.org/issues/processing/#worker

[46]   https://www.economist.com/special-report/2019/06/20/california-and-texas-have-differ-ent- visions-for-americas-future

[47]   Russell Baker, "Red Meat Decadence," New York Times, April 1973.

[48]   Up in the Old Hotel, Joseph Mitchell, 1992, "The Beef Steak." [49], https://amp.reddit.com/r/vegan/comments/3t0p1w/my_vegan_friend_claimed_that_vegans_are_less

[50]   http://www.silentlamb.org/CatholicSettles.htm.

[51]   https://www.livekindly.co/how-woody-harrelson-became-vegan-icon/amp/

[52]   Bible: (Genesis 1:29 NIV)

[53]   Struthious: A non-practical behavior of sticking one's head in a hole to avoid bad habits or problems, synonymous with ostriches.

[54]   "Proof of Vedic Cultures Global Existence" by Stephan Knapp. Chapter 6, "More

[55]   Sanskrit/Vedic Links with English Words and Western Culture."

[56]   Barbarian/Vedic: "Rewriting Indian History," by Fanny Gautier. Chapter 1, "A Look at Western Civilization to Understand the Vision We Have of India Today."

[57]   "Major World Religions from their Origins to the Present" by Lloyd Ridgeon.

[58]   "Violence Against Wives: A Case against the Patriarchy." New York. The Free Press, 1979: Do-bash Emerson, R., and Russel Dobash.

[59]   http://www.sas.upenn.edu/~psanday/books.htm/.

[60]   Jeremy Rifkin: "Beyond Beef," (92).

[61]   ibid

[62]   ibid

[63]   The American Heritage Dictionary, 2010.

[64]   http://www.all-creatures.org/

[65]   Charles Darwin: http://r.search.yahoo.com/_ylt=AwrSbD_q2hRWhoQAb6FXNyoA;_ylu=X-3oDMTEyMmdhZm50B   GNvbG8DZ3ExBHBvcwM2BHZ0aWQDQjAwMjNfMQRzZWMc3I/RV=2/RE=1444236139/RO=10/RU=http%3a%2f%2fwww.spaceandmotion.com%2fCharles-   Darwin-Theo-ry-Evolution.htm/RK=0/RS=U4SkLfV9O0o2lpofwJeA5Jb_SWQ-

[66]   The American Heritage Dictionary, 2010.

[67]   Indigenous People's Day - Wikipedia

[68]   "The Sexual Politics of Meat" by Carol Adams, 1990.

[69]   Jeremy Rifkin, "Beyond Beef," (92:247).

[70]   Genesis: (1:26).

[71]   Matthew 5:5.

[72]   Noah's Ark: http://christianity.about.com/od/biblestorysummaries/p/noahsarkflood.htm.

[73]   http://www.answers.com/topic/erasmus-darwin.

[74]   Merriam-Webster Dictionary defines the word Eurocentric as "centered on Europe or the Europeans; especially: reflecting a tendency to interpret the world in terms of western and especially European or Anglo-American values and experiences."

[75]   "God is Not Great," by Christopher Hitchens.

[76]   John Locke, Some Thoughts Concerning Reading and Study for Gentlemen in Wars (London, 1823), III, 296–97: Some Thoughts Concerning Education, IV, 182–84, and 11.

[77]   www.boisestate.edu/history/ncasner/hy210/tears.htm.

[78]   www.amazon.com/Ecotopia-Ernest-Callenbach/dp/0553348477.

[79]   Jeremy Rifkin, "Beyond Beef," (92: 79–80).

[80]   The Dorothy Dandridge Story, LMN (Life Time Movie Network), 2010.

[81]   Ku Klux Klan: http://www.civilwarbaptists.com

[82]   http://www.publiceye/larouche/Wohlforth.html.

[83]   http://www.inplainsite.org/html/dispensationalism.html.

[84]   Hal Lindsey, with C. C. Carlson, "The Late Great Planet Earth," (Grand Rapids, MI: Zondervan Publishing House, 1970).

[85]   Billy Graham, Approaching Hoofbeats: The Four Horsemen of the Apocalypse (Minneapolis, MN: Grason, 1983), pp. 222–224.

[86]   Johnson, Architects of Fear, pp. 28–29; F. H. Knelman, "Reagan, God and the Bomb"
, (Buffalo, NY: Prometheus Books, 1985); pp. 175–190; Boyer, "When Time Shall Be No More," p. 162.

[87]   Grace Halsell, "Prophecy and Politics: Militant Evangelists on the Road to Nuclear War," (Wesport, CT: Lawrence Hill, 1986).

[88]   http://www.publiceye.org/apocalyptic/Dances_with_Devils_2.html. [89] Genesis: (1:29, 30; 2:7, 19).

[90]   http://www.krishnajanmashtami.com/teachings-lord-krishna.html.

[91]   Genesis: (8: 21).

[92]   https://www.all-creatures.org/hr/what-00.htm

[93]   Ibid

[94]   Essenism: An ancient religion who opposes the slaughter of animals. A civilization revealed by the Dead Sea Scrolls. They lived an ascetic, monk like, communal existence in Palestine and Jerusalem before and during the time of Christ. Like Christ, they possessed no weapons or slaves.

[95]   Viatoris Ministries

[96]   Ibid

[97]   http://www.gotquestions.org/mercy-not-sacrifice.html [98] Amos: (5:21, 22, 24) (JB).

[99]   Isaiah 1:11, 13–16 (JB)

[100]   Hosea (2:18) (NEB)

[101]   Hebrews (4:13)

[102]   "The Lost Gospel of Thomas" by Steven Davies.

[103]   https://ivu.org/history/christian/clement.html

[104]   http://forums.catholic.com/showthread.php?t=574545

[105]   Jeremy Rifkin, "Beyond Beef," (92: 278).

[106]   live by the Sword Quote Matthew: (26:52).m

[107]   www.goveganradio.com/tag/rynnberry

[108]   http://www.responsibleeatingandliving.com/vegetarian-historian-rynn-berry-now-part-of- vegan-history

[109]   http://en.wikipedia.org/wiki/Albert_Schweitzer. (Refer to "The Relationship of Drug Addiction to Food Craving")

[110]   "Disciples: How Jewish Christianity Shaped Jesus and Shattered the Church," by Keith Akers,,   http://r.search.yahoo.com/_ylt=A0SO81PXrxNWjjIAvPxXNyoA;_ylu=X3oDMTEyZHQycWQ0BG NvbG8DZ3ExBHBvcwMxBHZ0aWQDQjAwMjNfMQRzZWMDc3I/RV=2/RE=1444159576/RO=10/ RU=http%3a%2f%2fwww.amazon.com%2fDisciples-Jewish- Christianity-Shaped- Shattered%2fd-p%2f1937002500/RK=0/RS=cDrPVTjKgHNP9JVDhcU2eycdzBY-

[111]   https://hippocratesinst.org/learning-centre/blog/archive/early-christian-vegetarian- co-munities/

[112]   St.Basilhttp://r.search.yahoo.com/_ylt=A0SO8ybUsxNWwrkAMFZXNyoA;_ylu=X3oDMTEy-dnJtaWo5BGNvbG8DZ3ExBHBvcwM0BHZ0aWQDQjAwMjNfMQRzZWMDc3I/RV=2/RE=1444160596/ RO=10/RU=http%3a%2f%2fforums.catholic.com%2fshowthread.php%3f t%3d574545/RK=0/RS=Zp-meOYc_KNz1yU8b0DVHbL44jlk-

[113]   St. Augustine http://r.search.yahoo.com/_ylt=AwrSbnbZthNWf2oAAdpXNyoA;_ylu=X-3oDMTEyMWk1N2o0BG   NvbG8DZ3ExBHBvcwMyBHZ0aWQDQjAwMjNfMQRzZWMDc3I/RV=2/ RE=1444161369/RO=10/RU=http%3a%2f%2fwww.brainyquote.com%2fquotes%2fauthors%2fs%2f-saint_augustine.html/RK=0/RS=VrAzVHwUkS5eIT55zuK.l1ziKkc-

[114]   The Bible: [Daniel 1:8], the Story of King Nebuchadnezzar [115]Peter http://r.search.yahoo.com/_ylt=A0SO8yFayxRWAIYAdjpXNyoA;_ylu=X3oDMTEyZHFtYXBlBG   NvbG8DZ3ExBHBvcwMz-BHZ0aWQDQjAwMjNfMQRzZWMDc3I/RV=2/RE=1444232155/RO=10/RU=http%3a%2f%2fwww.all-creatures.org%2farticles%2fearly.html/RK=0/RS=aA_BZKRXTrYiYm1GcxNs6PNUTgw-

[116]   St. Francis http://r.search.yahoo.com/_ylt=AwrSbnuGzRRWNXwA95xXNyoA;_ylu=X-3oDMTEydnJtaWo5B   GNvbG8DZ3ExBHBvcwM0BHZ0aWQDQjAwMjNfMQRzZWMDc3I/RV=2/ RE=1444232710/RO=10/RU=http%3a%2f%2fwww.ivu.org%2fhistory%2frenaissance%2fs t-francis. html/RK=0/RS=aF1ZROEkUv0wHhBfk9gsK5tZ82o-

[117]   St. Jerome http://r.search.yahoo.com/_ylt=A0SO8ob9zxRWVVUAfYBXNyoA;_ylu=X-3oDMTEyZHFtYXBlB   GNvbG8DZ3ExBHBvcwMzBHZ0aWQDQjAwMjNfMQRzZWMDc3I/RV=2/ RE=1444233341/RO=10/RU=http%3a%2f%2fwww.ivu.org%2fhistory%2fchristian%2fjer ome.html/ RK=0/RS=r8vFy1P0lKskJnFahMUC1TxIsCQ-

[118]   yogananda.srf.org

[119]   https://www.amazon.com/Lost-History-Christianity-Thousand-Year-Asia/dp/0061472816

[120]   The Ugly American: Traditionally, this epithet refers to perceptions of ostentatious, de-meaning, and ethnocentric behavior abroad by U.S. citizens. This term is also used to describe U.S. corporate and military protocol regarding draconian or jingoistic foreign policies.

[121]   http://www.shaolinkungfu.com/.

[122]   Popeye   http://r.search.yahoo.com/_ylt=AwrTHQ.d0hRWhysAix1XNyoA;_ylu=X-3oDMTEyZHFtYXBlBG   NvbG8DZ3ExBHBvcwMzBHZ0aWQDQjAwMjNfMQRzZWMDc3I/RV=2/ RE=1444234014/RO=10/RU=http%3a%2f%2fwww.youtube.com%2fwatch%3fv%3dnsVD spG8Ozl/ RK=0/RS=Ms0O4Glg_G_dHCM3bmi824EyHa0-

[123]   http://www.biography.com/articles/Albert-Einstein-9285408.

Much-needed introspection of our basic traditions and ethos will enlighten us to abstain from animal products and become a part of …

The Conscious Planet

CHAPTER 3

P.E.T.A. ORG.
People for the Ethical Treatment of Animals

Disclaimer*

Recently it has come to this author's attention that some vegans have become diametrically opposed to this organization due to issues related to euthanasia. But when an organization saves so many animals, they must also be responsible for all those animals' wellbeing for the rest of their lives. When there are so many animals to care for, it is inevitable that many will fall ill. This author would not like to be the one to make the decision as to which ones should be put down, but is it more compassionate to just let them suffer? These are moral questions we must all ask ourselves, and there is NO good answer, because each situation carries its own weight. This author does not believe that anyone from the PETA org. wants to see animals suffer.

This book could not have been written without mentioning Ingrid Newkirk, cofounder and president of PETA, along with all her compassionate colleges, and all the wonderful work they do to protect animals around the world. The People for the Ethical Treatment of Animals is the world's largest animal rights organization, with over 3 million members internationally. For more than 35 years, the PETA Org. has been making a difference toward animal rights while exercising true compassion for all living creatures.

PETA directs their activities to save the highest numbers of animals possible. Under scrutiny, are factory farms, laboratories, and clothing manufacturers (Especially the fur and leather industry.) PETA interacts with the community through animal rights legislation, public education, research, animal rescue, celebrity involvement, special events, and protest campaigns. PETA is known for their graphically provocative visual displays, e.g. naked models covered in fake blood, wrapped in plastic to simulate packaged meat. They are trying to convey a psychological impact on the public to make them understand the way our modern Western culture attempts to relegate and eschew its moral obligations toward the visceral nature of animal slaughter.

In Jan of 2010, PETA reported that the former First Lady (Michelle Obama), Oprah and Tyra Banks (3 very powerful, politically correct, and influential black women), all denounced the use

In the HBO special *"Death on a Factory Farm,"* 3/19/09, PETA hires an undercover agent to infiltrate and expose the cruel, morbid, and inhumane, factory pig farming industry. Featured in this documentary is undercover footage of pigs being put to death by hanging, while workers are standing around or going about their business, like this is a common place occurrence! The camera reveals the heartbreakingly inhumane, brazen and barbarous nature of the pork industry. In this HBO special, when the owner of one of these pork processing plants was subpoenaed to testify in court, while being under oath, he emphatically denied any acknowledgement that the animal may be suffering in an abysmal attempt to exonerate himself from this obvious act of inhumanity! His cavalier, brazen, and uncompassionate demeanor, displayed the "Pompous Pugnacity of a Pernicious Pork Purveyor!"

The Cargill Corp also displays irresponsible advertising with their bus which looks like a big cute pig. Why don't they show this HBO special to all those innocent kids before they load them on that bus and take them to a barbecue![2]

In addition, what this author also found disturbing during this HBO special was Ms. Newkirk's comment about her belief in atheism. This makes no sense, because Ingrid, you don't have to be a master in the science of metaphysics to understand that the true spirit of God exists in the eyes and hearts of all the wonderful animals that you and your organization saves. Bless you for all your ~~good~~ [God] work! [3]

Notes:

[1]   http://news.yahoo.com/s/ynews/ynews_ts1047
[2]   www.cargill.com
[3]   http://www.peta.org

All Lives Matter!

Animals are one of our closest links to the true spirit of God or creation. These creatures' minds and hearts are innocent to the corrupt and evil ways of modern civilization. They represent the purest and most unadulterated entities of living spirit.

So therefore…How can a civilized, spiritually evolved human being, without conscience or remorse, brutally murder in cold blood, one of these magnificent creatures of nature?

Before You Slaughter that Animal …

Realize They Have a Heart …

Where's Yours? …

Be Smart …

Look Deep into Their Eyes …

They Got Two Like You …

They Are the Mirrors to Their Soul …

Let the Truth be Told …

Their Right to Life …

Should Never be Sold !!!

By Displaying Kindness to Animals, We Automatically Teach Compassion to our Children, and Therefore We Become a Part of…

The Conscious Planet

CHAPTER 4

VEGAN MOVERS & SHAKERS

In the last chapter, we have already learned about Ingrid Newkirk, and all the wonderful work she does to help animals. Here is a list of some more of the most powerful and influential individuals in the world, that are also creating a vegan revolution! Veganism is the prerequisite for the future of humanity! These truly enlightened individuals are really helping to make a difference by bringing awareness to some of the most troubling issues we face into the future!

LEONARDO DiCAPRIO

Everyone knows that Leonardo DiCaprio, is a Mega Mogul superstar actor. However, many people may not be aware that he has done more to protect animals and save the environment, than almost any other individual in history! Besides endangered species, he's been on a worldwide quest to eradicate livestock production. In 2013, he donated 3 million dollars to the (WWF), World Wildlife Fund to help rescue tigers in Nepal. Later that year, through his foundation, he raised almost 39 million dollars in a charity auction, and in 2014, He became the executive producer of the documentaries'  "Cowspiracy," and also the Oscar nominated film "Virguna," which explains the plight of the great apes in Africa. In 2015, he donated an additional 15 million dollars for similar causes, working to combat climate change, prevent the extinction of animals and to protect the environment. In 2016, he was instrumental in the creation of the highly acclaimed documentary "Before the Flood," with executive producer, Martin Scorsese. Bravo Leo! [1]

Greta Thunberg

"At 16 years old, Swedish, environmental vegan activist, Greta Thunberg, was voted Time magazine's "Person of the Year" for 2019. She is the youngest recipient for this prestigious award of any other person in history. Greta represents a generational shift and growing worldwide movement of environmental awareness. Her impact has become so prolific, that all over the globe, from Hong Kong to Chile, people are demanding change! Since 2018, Greta has made incredible progress toward her vision of sustainability. Just in this short time, not only has she met with the Pope, addressed the heads of state at the U.N. and even debated with Donald Trump, but

she had also inspired 4 million people to join a climate change strike on September 20th 2019, which represented the largest strike of its kind in human history!"

Recently, in late 2019, at a climate change conference, Greta met with Academy Award winner and global warming champion, Leonardo DiCaprio. In an Instagram post, DiCaprio writes *"There are few times in history that voices are amplified at such pivotal moments and in such transformational ways – but Greta Thunberg has become a leader of our time! History will judge us for what we do today to help*

guarantee that future generations can enjoy the same livable planet that we have so clearly taken for granted. I hope Greta's message is a wakeup call to world leaders everywhere that the time for inaction is over. It is because of Greta, and young activists everywhere that I am optimistic about what the future holds. It was an honor to spend time with Greta. She has made a commitment to support one another, in hopes of securing a brighter future for our planet!" [2]

Woody Harrelson

Besides his iconic acting career, Woody Harrelson, has been an active proponent of Veganism for more than 30 years. Harrelson, being outspoken on this subject, has lectured and written to public officials on behalf of animal rights organizations, such as PETA, calling for legislation and pleading for public

awareness. He has also worked with vegan chefs, campaigning to get people to eat more plant-based foods.

Beyond his work with PETA, in 2019, Harrelson became involved in a monumental project, becoming a member of The Million Dollar Vegan campaign that works with world leaders on issues related to climate change, animal rights and world hunger. He even collaborated with a team of vegan celebrities (including Paul McCartney, fellow Vegan actors, Joaquin Phoenix and Evanna Lynch, along with plant-based physician, Dr. Neal Barnard), to meet with the Pope, in order to get him to go vegan. [3]

PAUL MCCARTNEY

Paul McCartney is one of the greatest Rock legends in history! His music has been quite prolific, stretching more than 6 decades! Today Paul is a serious vegan but was a vegetarian since 1975. He has devoted his life to animal welfare and supports such organizations as PETA, The Humane Society and the World Society for the Protection of Animals. In 2009, McCartney, along with his two daughters, launched the *"Meat Free Monday"* campaign.

Wherever he performs, McCartney strongly promotes veganism, and also makes an effort to discourage vendors from selling animal products.

> *"Many years ago, I was fishing, and I was reeling in the poor fish, I realized, I am killing him—all for the passing pleasure it brings me. And something inside me clicked. I realized as I watched him fight for breath, that his life was as important to him as mine to me!"*
>
> — (Paul McCartney) [4]

GARY YOUROFSKY

Yourofsky, is a living legend among the vegan community. He is probably the most prolific animal rights activist in history. Yourofsky has gone to jail many times over his activism. In 1997, Yourofsky, along with 4 members of the Animal Liberation Front (ALF), raided a fur farm in Blenheim, Ontario Canada, and released 1,542 mink (who were doomed to be killed for their fur), into the wild! This reportedly caused losses to this farm for more than a half million dollars! Subsequently, he was arrested and after being tried, was sentenced to 6 months in prison. After spending 77 days, he was finally released. Regarding his brazen experience of *"Caged confinement,"* Yourofsky stated, that it *"Reinforced my empathy and understanding of what these animals go through."* As of 2015, he had performed 2660 lectures to more than 60,000 People at 186 colleges and universities, in 30 states and several foreign countries.[5]

ERIC SCHMIDT

The term movers and shakers was invented just for people like Schmidt. This billionaire, Executive Chairmen of Google has prognosticated that *"A vegan revolution is coming!"* He states that the abstinence of animal products will be the *"Number one game changing trend of the future!"* Under his guise, in 2015, Google tried to buy the plant-based startup, Impossible Foods for a hefty amount of 300 million dollars, but the offer was rejected.[6]

JOAQUIN PHOENIX

He is a Golden Globe and Grammy award winning actor and was also nominated for 3 Academy Awards. In Late 2019 his movie debut of The Joker was a smash hit. Joaquin has been on a plant-based diet since he was 3 year old and has been active in animal rights awareness for many years. As was previously mentioned, he was one of the celebrities that went to visit the Pope, to try to make him go vegan. He is also an active member of PETA.

> *"Eating animals is absurd and barbaric"*
>
> — (Joaquin Phoenix) [6]

RUSSEL SIMMONS

With a net worth of 340 million (As of 2011), Simmons (Co-founder of Def Jam records), is one of the most influential entertainment moguls in history. Besides his Music career, he is also chairmen and CEO of

Rush Communications, film producer and has created the extremely popular lines of clothing; Phat Farm, Argyleculture and Tantris.

In 1997, this hip-hop icon made a life changing decision; he went vegan for reasons of health, compassion and environmental awareness. In 2015, after experiencing the powerful health benefits of a vegan diet, and while realizing the in-compassionate and environmental repercussions of animal agriculture, he published a book called *The Happy Vegan*.[8]

JOHN ROBBINS

John Robbins sacrificed an entire fortune, which would have been afforded to him as rightful heir to the Baskin-Robins Empire! He has disassociated himself with his family for over 2 decades! All this was done over moral reasons after witnessing the neglect and abuse of dairy cows. A milk producing cow can live up to 24 years old, however, due to heavy burden of modern industrialized milk production, a dairy cow will only live to 4 years old, and then ground into hamburger meat! So don't fool yourself, using dairy products is just a cruel as eating meat! Robbins has appeared on many television shows and documentaries, also being the author of 3 books; *"Diet For A New America,"* *"Food Revolution"* and *"The New Good Life."* [9]

DR. NEAL BARNARD

He was the Vegan Dr. that went with Leonardo DiCaprio and Woody Harrelson to visit the Pope. Barnard is the prolific author of more than 12 vegan and nutritionally related books with such titles like the *"Get Healthy, Go Vegan Cookbook,"* *"Foods That Cause You To Lose Weight,"* and *"Power Foods For the Brain,"* just to name a few.

> *"The beef industry has contributed to more American deaths than all the wars of this century, all natural disasters, and all automobile accidents combined!"*
>
> — (Dr. Neal Barnard) [10]

PATRIK BABOUMIAN

Patrik Baboumian, as a vegan, has set many world records for weightlifting, and at one time, was considered by many, to be the world's strongest man. However, if you ask him what being strong really means, he would tell you that *"True strength isn't about bulging biceps or how many pounds you can dead lift; it's about extending kindness, showing gentleness instead of cruelty, and exercising compassion!"* Baboumian breaks the typical stereotype of what a vegan is. Patrik published a book entitled *"VRebellion-1: How to Become a Vegan Badass!"* [11]

Notes:

[1]   https://animalfair.com/2015/07/16/leo-dicaprio-donate-15-million-to-protect-animals-and- conserve-the-environment/

[2]   https://blog.jetsettingmagazine.com/news/time-magazines-person-of-the-year-greta-thunberg

[3]   https://www.livekindly.co/how-woody-harrelson-became-vegan-icon/amp/

[4]   https://www.vegan.com/paul-mccartney/

[5]   https://en.wikipedia.org/wiki/Gary_Yourofsky

[6]   https://www.livekindly.co/googles-executive-chairman-predicts-vegan-revolution-coming/amp//

[7]   https://www.livekindly.co/joaquin-phoenix-vegan-hoodie-joker-premiere/amp/

[8]   https://rushcommunications.com/2018/06/russell-simmons-on-veganism/

[9]   https://www.miaminewtimes.com/restaurants/john-robbins-of-baskin-robbins-family-busts- dairy-industry-myths-and-more-6595994

[10]   https://en.m.wikipedia.org/wiki/Neal_D._Barnard

[11]   https://en.m.wikipedia.org/wiki/Patrik_Baboumian4

No army
can withstand
the power of
an idea
whose time
has come!

— VICTOR HUGO

(Based on the original version of this chapter, author Neil M. Pine, won an Eco Hero Award for a speech on non- sustainable practices, from the University of California at Riverside (UCR), on Earth Day, April 24, 2010.)

The Conscious Planet

CHAPTER 5

PEACE AND PROSPERITY

The Shocking History of the GOP (Bush, Reagan & Trump!)
(But Those Dems Aren't No Saints Either)

Special note:

REGARDING PRESIDENTIAL ELECTION 2020:

Anyone who claims that both parties have not been guilty in the past of rampant corruption and scandal is a "Lying Dog Face Pony Soldier!" This is not about right or left, but about what is clearly Right and Wrong! Under Biden, 200K illegal citizens flooding in a month; many bringing dangerous drugs and human trafficking while spreading Covid, and further exacerbating the problems with the homeless! I don't see the president mandating masks for these people! Furthermore, in a world going parabolic with violence, then do we really want to make it easy for terrorist and dangerous undocumented criminals to freely enter this country? The only possible benefit from such blatantly cavalier and dangerous policies, would be to help get the Democrats get reelected! This certainly doesn't help any U.S. citizens! There is also the scandal with Hunter Biden and Burisma, while Kamala Harris, launched her political career, based on the infidelity of former San Francisco mayor, Willie Brown!

This is why, for so many reasons, this author did not endorse either candidate and honestly, would be much too embarrassed to say that he did! However, for the sake of the environment and future generations, the American public was once again forced to begrudgingly pick between the lesser of two evils!

Greta Thunberg is not some Swedish fairy tale. This courageous young Vegan woman, represents a real movement, backed up by 11,000 climate scientists and over 6 million environmental protesters worldwide!

What we need is a third party which practices transparency and truly understands sustainability with compassionate ideals, that would outlaw factory farms and bolster alternative energy and transportation technologies, by taking subsidies away from dirty energy and retraining workers in order to help harness the clean energy of the future!

The Future of Humanity Hangs Over an Abyss!

GOP PLUTOCRACY:

A Legacy of Economic and Environmental Plundering "Corporations should be a servant of the people, not the master."

The legacy of the GOP, ("Greedy Old Politicians"), represents generations of "neoconservative" protocol, which has been conveniently hidden behind an evangelical shroud of eschatological dispensations, in order, to justify their long history of "scapegoating, demagoguery, Jingoism, hegemony, and egregious anthropogenic malfeasance"! (Refer to chapter 2. "The Psychology of the Cattle Culture.")

The Reagan, Bush, and Trump Administrations represented a classic example of corporate plutocracy, involving campaign funds, contributions, bribery, special interest groups, and government contracts. Corporations gain too much power through government funding.

Government supervision of corporate political funding should be initiated! New accountability standards must be established in order to restore integrity to corporate protocol. Corporate transparency regarding government funding and campaign contributions should be of paramount importance!

> "There can be no effective control of corporations while their political activity remains. To put an end to it will be neither a short nor an easy task, but it can be done."
>
> — (Theodore Roosevelt)

KENNEDY AND THE FEDERAL RESERVE:

During his presidency, John F. Kennedy realized what a serious threat to social mobility the Federal Reserve was and vowed to dismantle the banking system, thus trying to protect the integrity of the U.S. dollar. Many people feel that this may have led to his demise, rather than the Cuban Missile Crisis or the Mafia theory.

During the Nixon administration in 1971, with Kennedy safely out of the way, the Federal Reserve was empowered to establish their sinister plan to deregulate the U.S. dollar off the gold standard. Under the previous Johnson administration, paper dollars could only be printed to represent four times that of actual gold reserves which was fixed at $32 an ounce until it was deregulated in 1971. Subsequently from this point, the dollar collapsed and gold appreciated dramatically. By early 1981, it was up an astounding 2587%, reaching a record high at that time of $850 an ounce.

Thirty-four years previous to 1971, the money supply only grew at rate of less than twofold. However, thirty-four years after this heinous act against social mobility, the money supply had expanded thirteen times itself! No wonder, it's backed up by trees, not gold! Not only did the Federal Reserve issue bogus money to the American public after this deregulation, but they also got to *"keep all the gold"!* This deregulation created an enormous gap between the super wealthy and everyone else! [1] & [2]

Other financial scams by the powerful banking interests include the "tech bubble" of 2000 and the "subprime Real-Estate meltdown" of 2008. And now as of 2021, with Fed Treasurer, Jerome Powell, of the current Biden administration, the dollar may be set to collapse again, as the Fed attempts to print its way out of a "mountain of exponential debt!" They may likely fail; however, they did not directly create this problem, they inherited it! So where did all the money go? To "Capitalism"!

> "Bottom line today is the top 1% earns more than the bottom 50%. The top 1% owns more wealth than the bottom 90%. That gap between the very rich and everybody else is getting wider."
>
> — (Senator Bernie Sanders): [3]

So, with the current Biden Administration, how much different can it be? We have inherited the same banking system that prints endless amounts of money (Federal Reserve), which Kennedy tried to stop.

However, in order to try to deal with this 1% crisis, the Senate, on February 2, 2012, finally passed legislation which was specifically designed to avert government aids, administration officials, and all congressional members from using non-publicly disclosed information for purposes of conducting insider trading activity.

This aggressive move by the Senate was directly attributed to strong public sentiment, displaying a lack of confidence in government officials. Barack Obama proudly acknowledged and approved of this bill. This bill is referred to as the STOCK Act (Stop Trading on Congressional Knowledge). [4]

THE FORMER BUSH ADMINISTRATION:

A Scourge Against Humanity!

We remember the Holocaust because we don't want to see another Hitler. Yet, we still elected George W. Bush under the same false pretenses. Like Hitler, we must never forget George W Bush and his former administration for all their atrocities committed against nature and humanity!

All the deception, lies, and hypocrisy by the former Bush administration were unprecedented during any other administration in history! George W. Bush displayed beguiling intransigence in his quest and subsequent promulgation of Iraq; he couldn't be stopped. His actions were premeditated and based on erroneously contrived information! He demonstrated a pharisaical and demagogic domination of the political arena by cheating Al Gore out of the votes, lying about WMDs, and creating 9-11 along with an illegal energy war!

Throughout history, wars have been waged based on the struggle over energy and other natural resources. The provision of Petro-energy resources to the United States and other emerging global economies creates international tension. It's hard to search the globe for oil without infringing upon other people's cultures and political agendas. The development of renewable energy will mitigate our demand for international energy resources as well as support a new green economy.

Socio-Economic Benefits of Sustainability:

1.) Autarky:

The Webster's Dictionary defines the word autarky:

1. "Self Sufficiency, Independence; Specifically: National Economic Self-Sufficiency and Independence."

2. "A Policy of Establishing a Self-Sufficient and Independent National Economy"

This state of autarky would be characterized by our emerging green economy. Developing independent sources of green energy breaks our dependence on foreign oil, as well as the monopolistic control of pretentiously corrupt utility companies such as Enron. It gives the "power back to the people," thus preventing corruption, war, pollution, and global warming. And in turn will help to strengthen our economy and give back a new sense of "pride" to America. [5]

2.) Patriotism:

This new "pride" has real patriotic implications: By making a conscientious effort to utilize alternative energy—buy electric cars, solar panels, etc.—this shows true patriotism. Doing good things for your country is not fighting in a foreign war to protect the oil rights of huge multinational corporations who don't necessarily have the United States in their best interest!

The Webster's Dictionary defines the word patriot: "a person who loves their country and supports its interests." To do what is right for your fellow man. If people truly love their country, then they would do what's necessary to stop the wanton destruction of our environment and the senseless killing of innocent Americans in Iraq or other foreign countries. Our founding fathers intended patriotism to be "for the people, by the people," not "for the multinational corporations, by the government," like George Bush would have had you believe. I would call this is real patriotism rather than to "succumb to the nefarious whims of a supercilious, megalomaniacal, second-generation, war-mongering oil barren"!

George Bush was not just "supercilious," but he was also a "super-silly-ass!" Because someone should have warned him and his entire Klan that "no army can withstand the power of an idea whose time has come." This magnanimous quote by prolific author Victor Hugo gives strong affirmation to this "patriotic sustainable paradigm." [6]

George Bush, in 2007, gave a speech about his justification for increased war spending in front of the Texas Cattleman's Association. This is the same organization that unsuccessfully tried to sue Oprah when she publicly stated on her show that she would "never eat another burger." It's another case of the blind leading the blind!" Bush represented war and oil, and the cattlemen represent beef: three industries which have no sustainable future! These "good old boys" displayed an antediluvian mentality that, like dinosaurs, will someday become extinct! They represented the antithesis of green anthropogenic awareness by engaging in dirty, filthy, cruel, uncompassionate, and non-sustainable practices. Let's just hope that people see through their greed and ignorance before it's too late! Just by making a conscious effort to change our lifestyle, we can save ourselves from the cataclysmic ecological destruction created by the overproduction of livestock and the use of fossil fuels, which also fuels the U.S. war machine! [7]

So, ask yourself this question; is it patriotic to jeopardize the lives of young Americans so that big international corporations can steal oil, report record earnings, thus use all this wealth and power to continue keeping us dependent on foreign oil while making the taxpayers (you and me) pay for their protection and at the same time raising the cost of our fuel to the very people who risked their lives and paid for this protection? If this is not an "insult to injury," then this author doesn't know what is!

With over a trillion dollars spent militarily (estimated at 1.7 trillion as of November 2007)— paying for soldiers, weapons, food, transportation, housing, and general war infrastructure—all of this money, energy, and human resources could have been utilized to develop safe, clean alternative sources of energy, therefore breaking our dependence on foreign oil, saving the lives of young Americans and finally stopping pollution and the production of greenhouse gasses!

By eliminating our dependence on foreign oil provided to us by huge multinational oil conglomerates, we will spur a new green economy bolstered by technology. It will provide great opportunities to small green contractors, businesses, and fledgling corporations. When it comes to providing safe, clean alternative energy and fuel to Americans by Americans, jobs will not be outsourced. We no longer have to subjugate and oppress people of third world nations. A new entrepreneurial spirit will prevail, and a pride in knowing that we're no longer at the mercy of draconian foreign policies created by big oil and the U.S. war machine!

By patronizing foreign oil as consumers, we fuel the fire of hatred toward Americans, as we make ourselves dependent on the very people that we label as terrorists, while at the same time financing their operations against us! Without a doubt, a percentage of every dollar spent on gas inevitably ends up in their pockets!

Mankind must put an end to war or war will put an end to mankind.

— (JFK)

Bush the Barbarian's (the New Genghis Khan) Iraqi occupation incubated one of the greatest scourges against humanity in modern history! (Refer to chapter "The Psychology of the Cattle Culture.")

The Most Honest Public Statement Ever Made by G. W. Bush:

"Our enemies are innovative and resourceful, and so are we. They never stop thinking about new ways to harm our country and our people, and neither do we!"

— (George W. Bush, Washington, D.C., Aug. 5, 2004)

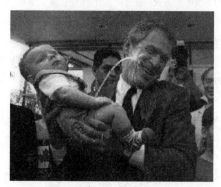

"Oops, the truth slips out"

Bush and Bin Laden Family Dynasties:

There can be no denial that the Bush family has carried on an intimate and protracted business relationship with the Bin Laden family and other Saudi elite for more than thirty years! In a book by Craig Unger entitled *House of Bush, House of Saud: The Secret Relationship between the World's Two Most Powerful Dynasties,* he describes this thirty-year history of business interactions between these two powerful families.

This book was banned in England, dropped by the British publisher for fear of Saudi lawsuits. Starting in the '70s, Saudi Arabian oil sheiks were seeking military protection, political influence, and investment opportunities from the United States. Some startling conclusions were reached in this book through interviews with three former CIA directors, top Saudi and Israeli intelligence officials, and over one hundred other credible sources of information.

In Houston, Texas, during the '70s, George W. Bush acquired his first oil company with funding coming directly from the super wealthy and influential Bin Laden family. On the day of 9-11, George Bush Sr. was at a business meeting at the Ritz Carlton Hotel in Washington, D.C., with one of Osama Bin Laden's brothers. [8]

In his book *See No Evil*, Robert Baer, a twenty-one-year veteran of the CIA, most of that time spent in the Middle East, extrapolates about this "intimate" relationship between the Bush family and the Saudi elite which he feels led to the intelligence failure prior to the WTC attack. [9]

9-11 Congressional Report: In the much-anticipated congressional hearings directed toward this intelligence failure in regard to 9-11, an eight hundred-page report was compiled in the summer of 2003. In this report, twenty-eight pages were blacked out to the public, ordered classified. This report focused on what each government agency knew, or should have known, prior to this disaster. It has been speculated that the twenty-eight missing pages of this document directly link the 9-11 hijackers to the Saudi government. [10]

On September 23, 2001, Colin Powell promised to present evidence to the American public which would acknowledge and document Bin Laden's role in 9-11. This evidence never surfaced. [11] & [12]

Rather than to be interrogated, held without bail, water-boarded, and tortured like the other political prisoners at Guantanamo Bay, Osama Bin Laden's family and relatives were all safely flown out of the country on an executive U.S. military jet two days after 9-11. [13]

> "Dick Cheney did more to encourage terrorism against Americans than almost any other official in U.S. history!"
> — (Jeremy Scahill), 6/5/09, author of *Blackwater: The Rise of the World's Most Powerful Mercenary Army*, 2007 [14]

Cheney and 9-11: A Paradox in Sanity or Insanity?

The Bush administration manufactured a crisis which enabled naïve and credulous Americans to justify a warlike mentality!

> "9-11 made possible what Dick Cheney called 'the war that won't end in our lifetimes.' This is a war that in reality is an energy war and 9-11 was its pretext."
> — (Michael Kane, staff writer for From the Wilderness.com). [15] & [16]

Michael C. Ruppert, in his book *Crossing the Rubicon: The Decline of the American Empire at the End of an Age of Oil*, quotes several key points which give conclusive evidence of a Bush administration–9/11 connection. In the first of which he states *"I will name Vice President Richard Cheney as the prime suspect in the mass murders of 9/11 and will establish that, not only was he a planner in the attack, but also that on the day of the attacks he was running a completely separate command, control, and, Communications system which was superseding any orders being issued by the FAA, the Pentagon, or the Whitehouse Situation Room."*

— (Michael C. Ruppert), *Crossing the Rubicon: The Decline of the American Empire at the End of an Age of Oil* [17]

Naomi Klein, author of *The Shock Doctrine: The Rise of Disaster Capitalism*, outlines the jingoistic nature of U.S. foreign policy, especially criticizing the Bush Administration's handling of Iraq, taking advantage of political chaos in order to profit from war. *"War prosperity falsely drives up GDP."* The author explains the concept of *"disaster capitalism,"* which entails a description of *"how the global 'free market' has exploited crises, violence, and shock over the past 3 decades to promote radical privatization that benefits large corporations and powerful special interest groups."* [18]

The American Civil Liberties Union had been trying to increase public awareness of human rights violations with their "close the Guantanamo Bay" prison campaign. While answering a question regarding prisoners' rights and inhumane treatment at Guantanamo Bay, Vice President Dick Cheney commented that the prisoners were being treated great. *"It's just like they're at a tropical resort"* [19]

Like a tropical resort huh?

"So come on down to Dick Cheney's 'Guantanamo Bay Resort Hotel' Now forget surfing, we got waterboarding. Tired of Bingo? We play ass pyramid.

So come join the 'jihad for lunch bunch,' where you cannot only get a 'side of hummus,' but you can also get a 'homicide!"

Also, by creating 9-11, the Bush Administration was able to successfully circumvent its involvement with Enron by creating the ultimate distraction.* They covered up their relationship in the Enron scandal by making themselves into heroes instead of the heathens that they actually were, while at the same time brainwashing the American public in to justifying a war that was totally unjustifiable! The public was so outraged, terrified, and perplexed over the state of affairs, that they were totally blindsided by all this government propaganda. [20]

In 2002, Senator Ted Kennedy pleaded to congress, warning about the Bush Administration's erroneous claims about WMDs and further stated aggressive military action could only exacerbate the already-delicate relationship between the United States and the Middle East. He clearly said that there was a manipulation of intelligence to make a case for war, with no solid evidence to substantiate such action. [21]

The Pentagon at that time was pressuring political analysts to go along with the Bush administration's propaganda, therefore loosely using the *"war on terror"* as a motivational theme to invade Iraq. Even going as far as to give erroneous information to top cabinet officials such as Colin Powell, who aggressively remonstrated his belief regarding this position to the United Nations, that Iraq was a major threat based on the false testimony provided to him!

Subsequently, after finding out that he had been duped, Colin Powell quietly resigned to save whatever little integrity and dignity he had left! [22]

On February 27, 2003, the U.S. diplomat, John Brady Kiesling, resigned, citing a *"distortion of intelligence"* and *"systematic manipulation of American opinion."* [23]

Even former Fed Chairman, Alan Greenspan, in 2007, to the chagrin of the Bush administration publicly stated, *"I'm saddened that it is politically correct to acknowledge what everybody knows: The Iraq war is largely about oil."* Defense secretary at the time, Robert Gates was quick to denounce Greenspan's statement; however, this was the basis for Greenspan's new book at that time, *The Age of Turbulence*, which contended that the motivating force for the war was to *"maintain U.S. access to the rich oil supplies of Iraq."* [24]

On September 6, 2001, former CEO of Enron, Jeff Skilling, tried to sell two hundred thousand shares of Enron. On September 7, 2001, Fed Funds Futures (a hedging mechanism for interest bearing instruments), made new contract highs. This action led to an inverse effect over interest rates, which went sharply lower. September 7 fell on a Friday, so this was the next to last trading day before 9-11. Monday was the tenth and Tuesday was 9-11, respectively. This evidence suggests that someone in the Federal Reserve knew about the coming attacks. There was no coincidence that Skilling was a member of the Houston branch of the Federal Reserve Board! [25]

5/23/02: Subpoena voted in by Senate subcommittee questions the Bush Administration's relationship to Enron. Cheney and Bush both claim *"executive privilege,"* refusing to answer any questions and therefore effectively placing themselves *"above the law."*

> *"Secrecy is a favored tool of the imperial presidency, and the Bush administration's stonewalling on its Enron connections signals that its declaring war on openness and is bent on quashing this scandal by any means. How about, for instance, distracting us with an endless war on an "axis of evil"?*

— (The Nation.Com) [26]

President Eisenhower, in one of the greatest speeches in American history, during his farewell address, emphatically warned the American people to beware of unchecked power in the military-industrial complex.

> *"This conjunction of an immense military establishment and large arms industry is new in the American experience ...*
>
> *In the councils of the government, we must guard against the acquisition of unwarranted influence, whether sought or unsought, by the Military-Industrial Complex. The potential for the disastrous rise of misplaced power exists and will persist. We must never let the weight of this combination endanger our liberties of democratic process."*

— (Dwight Eisenhower, Farewell Speech to the Nation,

Jan. 17, 1961) [27]

His daughter's statement from an appearance on Bill Maher television:

> *"In the original draft of President Eisenhower's speech, it was "Military Industrial-Congressional Complex." And the congressional part was taken out because the president felt that he'd had excellent relations with a democratic congress and didn't want to get into name calling on his way out."*

— (Susan Eisenhower) [28]

Some of the other guest opinions:

"When the military says, "We don't want this," and congress is trying to shove this down their throat, that's a sad statement of the political system that we have right now."

— (Anthony Woods) [29]

"Until we start to see some of that corruption actually prosecuted and some of these people start going to jail, the Military-Industrial complex is going to keep jamming weapons systems at the Pentagon, force-feeding them, that the country doesn't need and the country doesn't want!"

— (John Heilemann, political author) [30]

As of 2021, these aforementioned factors had become even more relevant, as powerfully pretentious defense contractors are pushing in Capital Hill to create legislation to make drones and other heavy military equipment (such as tanks, as was witnessed in Ferguson Missouri), mandatory requirement by Federal, State, and local municipalities; and not just for destroying our privacy through surveillance, but they also want to actually be able to use these technologies (just like for terrorists), to seek, target, and destroy U.S. citizens, when "THEY" would deem it necessary. Of course, the Mainstream Media would love to broadcast such mayhem to increase their ratings, thereby, at the same time, further perpetuating this technology simply by getting everyone to watch it! [31]

Correlation between the Bush Administration and the Third Reich:

The Reichstag Fire and the World Trade Center:

One of the key points in Eisenhower's speech was to *"endanger our liberties of democratic process."* By giving too much power to the *"military industrial complex,"* we create a greater propensity toward imperialism which lays the foundation for a military dictatorship. There are striking similarities to the Reichstag Fire of 1933 and the WTC disaster. Both of these events were insidiously orchestrated by unpopular minority leaders to squelch the opposition. Similar to actions taken by the Nazi regime, the Bush Administration played upon people's fears with 9-11. In the same manner, the **Nazi** regime in 1933 burnt down their own federal building, thus declaring martial law and creating a police state. [32]

> *"Whether it is a democracy, or a fascist dictatorship, or a parliament, or a communist dictatorship; voice or no voice, the people can always be brought to the bidding of the leaders. That is easy. All you have to do is tell them they are being attacked, and denounce the peace makers for a lack of patriotism and exposing the country to danger. It works the same in any country."*

> — (Hermann Goering, Hitler's deputy)

> *"Terrorism is the best political weapon for nothing drives people harder than a fear of sudden death ."*

> — (Adolf Hitler)

Not surprising, George Bush's approval rating dramatically increased after 9-11 from a 50% approval in early September of 2001 to almost a 90% approval rating by October of 2001. [33]

Both events were contrived and physically could not have happened the way they were reported, and both resulted in what was termed *"a temporary suspension of civil liberties"* in order to deal with a terrorist threat.

The Enabling Act of 1933, enacted by Hitler's regime, passed in regard to the Reichstag Fire, which was supposed to only "temporarily" suspend civil liberties in order to deal with communists. This so-called temporary statute of emergency lasted for twelve years until Germany lay in ruins! [34]

9/25/01: Justice Department lawyer John Yoo declares a *"flexibility statute"* on the Fourth Amendment, stating that *"the government may be justified in taking measures which in less troubled conditions could be seen as infringe-*

ments of individual liberties" Yoo subsequently creates the *"doctrine of pre-emption,"* inferring that the Bush administration, during times of war, may act against groups or individuals even if it would be *"difficult to establish that they have been or may be implicated in attacks."* [35]

10/25/01: The Patriot Act Passes 98-1 in the Senate. This bill was designed to impugn and compromise the very foundation of the U.S. Constitution. Passed shortly after 9-11, and was intended to *"temporarily"* suspend civil liberties to deal with the threat of Islamic extremist. [36]

In early 2002, the Bush Administration approves of "the program," similar to Hitler's Enabling Act, which further threatens the constitutional rights of the public by allowing the National Security Agency (NSA) to maintain surveillance of U.S. citizens without a warrant, court approval, or a sign off from the Justice Department! 12/22/02: (This was written 2 years before anyone heard of Christopher Snowden.)

> *"You said we're headed to a war in Iraq. I don't know why you say that ... I'm the person who gets to decide not you!"*
>
> — (Bush to the Press Corps)

> *"It's the Law that governs and not the man."*
>
> — (Thomas Paine, *The Rights of Man*)

Another parallel between the Bush family and the Third Reich was George Bush's "Quisling," Nazi-supporting grandfather, Senator Prescott Bush, who, contrary to popular belief, actually made his fortune helping to finance Hitler's regime rather than from the oil business! His bank was seized by the U.S. government, and he subsequently was charged with a war crime for aiding and abetting the enemy! [37]

Hitler & Grandpa Bush

The Nazi regime violated Geneva Convention standards with the creation of concentration camps. In a similar fashion, Cheney and the Bush Administration created Guantanamo Bay, which, according to international law set forth by the United Nations, also illegally detained and mistreated political prisoners.

Both Bush and Hitler were elected under questionable circumstances, using strikingly similar tactics involving propaganda, demagoguery, and ballot fixing!

Also in sympathy with Hitler's *Modus of Operandi*, was Bush's prolific record as the most callous Governor in U.S. history, subsequently earning the title as the "Texecusioner!," executing 131 prisoners during his 5 year tenure as Governor! We can all rest assured, that he would have made Hitler and his grandfather very proud! [38]

And just like Hitler, George Bush was originally elected in the same way that Saddam and many of Bush's fellow Texans were condemned: "under false pretenses!"

In the aforementioned, 60 Minutes special report, (From chapter 2) on 1/28/10, revealed that now-retired Supreme Court Justice, John Paul Stevens, stated that he believes that throughout his career, one of the greatest constitutional injustices in history was the primary election of 2000, where Al Gore was cheated out of the vote by house majority rule, which mandated that a recount would not be necessary. Yet at that time, Stevens expostulated his understanding to the Senate Judiciary Committee that there was no constitutional basis in order to justify house majority rule! He saw no reason to not demand a recount and felt that something was terribly wrong with the system at that time. To this day, he still feels very guilty about the outcome, stating *"The entire history of the world would have been different, if we only could have initiated that recount!"* [39]

According to the book by Pulitzer Prize–winning journalist Ron Suskind, *The One Percent Doctrine,* which was amassed mainly from CIA insider sources, claims that the video tape of Bin Laden, released four days before Bush's reelection on October 29, 2004, gave strong supportive evidence that Bin Laden's diatribe, actually helped Bush to get reelected! [40]

Also, in a mind-boggling disclosure by FBI sources, the FBI actually filed a document on Osama Bin Laden's behalf which protected his anonymity. This document is referred to as an "Exemption 6," which gives them the right to withhold all personal information including medical files, therefore giving Osama Bin Laden invasion of privacy rights. Ironically, wasn't the Patriot Act designed to supersede these rights? [41]

Hitler initiated the "propaganda ministry." The Bush Administration effectively utilized the "Mainstream Media" (MSM) through obnoxious, overbearing, and even racist personalities such as Rush Limbaugh, Sarah Palin, Glen Beck, Ann Coulter, and Bill O'Rielly. The American public was being "brainwashed" so bad that even prolific media icons such as Phil Donahue (who had been on the air for thirty years), were forced to resign after being labeled as "unpatriotic" under the onslaught of "pro-Bush demagoguery." As was stated by the Democratic Underground dot com, *"When MSNBC fired Phil Donahue we had no other anti-war voices."* [42]

The book *Mien Kampf (My Struggle)* was written by Hitler while in prison and spoke about his megalomaniacal goal to take over the world, which included the genocide of Jews by using poison gas. Most people during this time dismissed this work as the ravings of a madman.

In a similar note, the "Project for a New American Century" represented a think tank comprised of "neoconservatives" who controlled "key leadership positions" under the Bush/Cheney regime. In their report entitled "Rebuilding Americas Defenses" in September 2000, which was endorsed by Cheney, Jeb Bush, Donald Rumsfeld, Wolfowitz, and many others; it prognosticated that *"their goals of world domination would be difficult unless there was a new "Pearl Harbor."* Was 9-11 this new "Pearl Harbor" they were referring to? And what's even more disturbing, in light of recent evidence about the origin of Covid-19, is that they also noted in their report, that genetically engineered biological weapons *"may transform biological warfare from the realm of terror to a politically useful tool."* Did this PNAC report represent a *"new millennium Mien Kompf"*? [43]

After his illegal occupation of Iraq, Bush made a national TV announcement which should be construed as reminiscent of Third Reich protocol. He stated, *"We didn't start a war in Iraq, we freed the people."* And what he said next was even more disturbing, *"And we are going to free people all over the world wherever democracy is threatened."* Therefore, if another country has a commodity that the United States wants, then this gives them the right to go in and steal it, while at the same time killing a major portion of their population. Bush calls this "killing and plundering" freeing the people, and he wants to free people ("kill and plunder"), all over the world

wherever democracy is threatened. According to Bush's theory, any country who doesn't want to give away their natural resources or do business with the United States is a threat to democracy and would therefore be subject to invasion! Then doesn't this mean that Hitler was also very Democratic? [44]

Also, for eight years, George Bush, bankrupted our capitalistic system, giving all the tax credits away to his *"good oil boys"* who raked in record profits, while at the same time, taxing the American public and borrowing over a trillion dollars from foreign countries to fight a war that only benefited the oil companies. Historically, the GOP has displayed a legacy of these egregious environmental policies:

2001:

In 2001, the Bush Administration released "The National Energy Policy Report" (NEPR). President Bush appointed Dick Cheney, former CFO of Halliburton Oil before he took office as vice president. However, he was still officially on the Halliburton payroll and maintained a securities position in the company of 430,000 shares. The NEPR report was based on recommendations by major coal, oil, and nuclear interests, most of which being major Bush presidential campaign contributors. There were secret meetings between Cheney and major oil and gas companies, including BP, the National Mining Association, and the American Petroleum Institute. In this ecologically deplorable plan, only 7 of the 105 recommendations included renewable energy! Ironically, they proposed to fund these clean energy projects by opening up the Arctic National Wildlife Refuge (ANWR) for oil drilling, also earmarking 1.2 billion dollars in bid bonuses from the leases of the (ANWR.)

2002:

On 2002, the Bush administration released their fiscal budget cuts, which included a 33% reduction in solar and renewable energy, also severely cutting the R&D in these industries. In 2002, the House Energy Bill included a 33.5-billion-dollar tax break for *"dirty energy."*

2003:

The GOP-run congress backed up the House Energy Bill which entailed 23.5 billion dollars in tax breaks for major energy companies.

2004:

Once again in 2004, the Bush Administration cut another 25 million dollars from the renewable energy budget which would have been used for solar, wind, biomass, and geothermal applications.

2005:

In 2005, the Department of Energy cut energy efficiency and renewable energy programs by 50 million dollars, which was equivalent to a 4% cut. Also in that year, amendments in the house failed to raise fuel efficiency standards for automobiles. The Interior Department's Minerals Management Services, the agency responsible for managing oil and gas resources on the outer continental shelf and collecting royalties from the companies, made a decision to let oil companies determine the severity of environmental impact without government intervention. In essence, these stipulations allowed the oil and gas industry to regulate itself!

2006:

A bill passed by the House of Representatives allowed drilling for oil in the Arctic Refuge. They also passed the American-Made Energy and Good Jobs Act that same year, which allowed leases of coastal strips of the Arctic National Wildlife Refuge, which encompasses an area of 1.5 million acres. In 2006, there were even more proposed renewable energy cuts, with the elimination of geothermal funding and reductions of conservation funding by 6.34%.

2007:

In 2007, there was a legal loophole which allowed major oil and gas producers to default on their payments of 865 million dollars in taxes, while at the same time a 16% budget cut for renewable energy R&D was taking place! The Bush Administration had publicly stated that they were opposed to the expansion of renewable energy! President Bush went as far as to threaten to veto the Energy Independence and Security Act, which included a renewable electricity standard and tax cuts funded by the elimination of many tax subsidies for major oil producers, adding up to 13 billion dollars!

2008:

In 2008, there was another proposal to cut renewable energy programs by 27%. Also in that same year, the Bush Administration directly opposed the passage of the House Bill to Renewable Energy and Energy Conservation Tax Act (HR 5351). In addition, during his last year in office, Bush lifted the moratorium on offshore drilling, hence the slogan "Drill Baby Drill," which subsequently led to the BP disaster![45]

Anti-Environmental Policies of the Former Reagan Administration

In sympathy with the ludicrous statements made by Trump in 2019, about nuking hurricanes and windmills causing cancer, Ronald Reagan, while addressing questions regarding conservation issues during his California guber-natorial election of 1966, was also noted for making some ignorant and callous remarks; *"If you've seen one redwood tree, you've seen them all,"* and *"Trees cause more pollution than*  *automobiles."* This anthropogenic negligence was further exemplified by Reagan's presidential appointments of James Watt and Ann Gorsuch to Department of the Interior, and the head of the EPA, respectively.

> *"Never has America seen two more intensely controversial and blatantly anti- environmental political appointee's than Watt and Gorsuch."*
> — (Greg Wetstone, director of Advocacy of the National Resources Defense Council, who served on the hill during the Reagan era as chief of Environmental Council at the House Energy and Commerce Committee)

Under the Reagan Administration, Watt and Gorsuch, aggressively attempted to undermine and circumvent environmental policies established by "asphyxiating" the Clean Air Act with their proposals to weaken air pollution standards and to "taint" the Clean Water Act by lowering limits on water pollution control. Their political infidelity was characterized by this egregious environmental negligence. Watt was later fired, and Gorsuch was forced to resign over major scandal and corruption in her agency revealed by congress.

When former Secretary of the Interior James Watt was questioned about his blatant disregard for the environment and how it may affect future generations, he infamously stated, *"We don't have to protect the environment, the second coming is at hand."* Did this mean that the **Reagan Administration** no longer wanted Americans to plan for their children's future? Is this the all-American dream our founding fathers would have envisioned? *"C'mon' honey, let's hit Vegas with the kids' college fund, Ronnie said it was OK."*

> *"The Reagan Administration adopted an extraordinarily aggressive policy of issuing leases for oil, gas, and coal development on tens of millions of acres of national lands—more than any other administration in history."*
>
> — (David Alberswerth, Wilderness Society) [46], [47]

THE TRUMP ADMINISTRATION

Trump was elected primarily due to the public being weary about traditional politicians and it was hoped that his independently wealthy status would help to preclude him from acquiescing with special interest groups. However, in such a sensitive time in history, in terms of political correctness, Trump initially picked out the most blatantly prejudiced and pro big pharma, anti- marijuana, Attorney General, Jeff Sessions,  and a completely anti-environmental Secretary of State. Rex Tillerson, who was formally the CEO of Exxon Corporation.

The Dinosaurs of the Former Trump Administration

In today's world, we need to display more racial unity, not hate. We needed a sustainable visionary like Elon Musk, more than ever, rather than some old dinosaur like "Tyrant-a-Sore- Ass-Rex" Tillerson!

This administration had already displayed racial and sexual improprieties, compounded with gross malfeasance regarding the environment, human rights and even marijuana. People will not stand for this anymore. There is no place for such behavior in a so called modern and civilized political system, as these actions can only destroy unity within the party as well as intimidate the American public!

How and why Trump was so naïve, as to surround himself with so blatantly, politically incorrect figures, is beyond comprehension. This author is neither a Democrat nor a Republican, but this is not about being "Right or Left, but about what is clearly Right and Wrong!"

Former U.S. Attorney General, Jeff Sessions, was quoted to have said *"I used to think the KKK was okay until I found out they were smoking pot!"* So, lynching people and burning down churches is okay, but smoking marijuana is where he draws the line? This guy was prehistoric for sure, or was he merely protecting big pharma? In a similar note, Trump also put his foot in his mouth when he referred to Haiti as a *"Shithole"* nation, stating that *"Why can't they be more like Norway?"*

If Session's political mindset wasn't already backward enough, then Trump really outdid himself by appointing another dinosaur like Tillerson! In doing so, Trump has made the U.S. the world's largest producer of oil. However, Trump was also a bit of a dinosaur himself for not realizing the implications of his own folly. As technology advances, the propensity for the obsolescence of crude oil will someday come to fruition! We will inevitably end up over invested in an antediluvian energy infrastructure. We must make a major move in alternative energy NOW, if we wish to stay ahead of the geopolitical, economic curve. This is not only of paramount importance to the future U.S. economy. but is also critical toward the survival of the planet! Tillerson encouraged subsidies for his cronies' oil companies rather than to support sustainable protocol! [48]

Besides appointing a visionary like Elon Musk to Secretary of State, if Trump would have really wanted to get reelected, then he should have appointed a black, female Attorney General, who is also pro marijuana.

(Activist in dinosaur costumes protested Rex Tillerson's appointment as the U.S. Secretary of State) [49]

Trump and the Deep State:

The "Deep State" represents what is referred to as a shadow government. It is theorized to be an organized effort between unelected government bureaucrats, both liberal and conservative, in order to undermine presidential protocol. The Deep State is an intricate maze, similar to a block chain of subversive government officials. Secrets ubiquitous, yet impenetrable! Whistleblower, Thomas Drake, states that in regards to the Deep State, the NSA's old joke was that

"Presidents come and go, but we're still here!" And whether he realized it or not, Trump was playing into the hands of the Deep State with his bloated 2.2 trillion-dollar military budget! Now like a spoiled child, he has created "Space Command!" He could not have been playing into the hands of the Deep State any better; $600 dollar toilet seats and the cost of just 1 screw could be hundreds of dollars! He was making billions for defense contractors! Hundreds for just one screw? Once again, the American public is getting screwed! Whether you agree with Bernie Sanders or not, you must admit, that all this profligate military spending could have paid for a lot of health care and education [50]

Here comes this pompous reality star who's going to turn Washington upside down, and after decades of established underlying control, he still thinks he's going to change things overnight? JFK was also charismatic and from a wealthy family, but the Deep State squashed him and his brother like a bug! However, Trump is not that stupid; he knows his boundries, but to be accepted by the current political infrastructure, he must therefore, acquiesce to a certain degree, as he's dealing with powerful multinational corporations and volatile world governments. At any time, his actions could potentially trigger a world war or cost these powerful corporations billions in profit!

In an interview, in 2019, with the CEO of Ali Baba, Jack Ma, he stated that the U.S. is frivolously wasting trillions of dollars on military spending that should have gone for alternative energy infrastructure. [51]

Trump and the Endangered Species Act

The Endangered Species Act, originally enacted in 1973, was intended to protect both the fauna and flora. Just look at the rainforests burning out of control in late 2019. We are destroying beautiful pristine jungles and critically delicate animal habitats in the process. This is an extremely serious problem that this author has been trying to bring to people's attention for almost 40 years! (Refer to Endangered Species section)

In 2018, a scientific journal was published by the Intergovernmental Science Policy Platform on Biodiversity and Ecosystems. In this report, it is claimed that up to one million animal and plant species could face extinction by 2050 [52]

Trump's environmental policies were atrocious (endangered species, oil drilling, air and water pollution, etc…) He acquiesced with the protocol of major multinational corporations. These cutbacks in environmental standards were reminiscent of the Reagan and Bush Administrations.

> "Without the president's acknowledgment of climate change as a threat to our economy, our environment, and our health, his record on the environment can only be described as a total failure"
>
> "Under this president we've seen more than 80 rollbacks of public health and environmental protections on everything from the clean Power Plan to pesticides. Enforcement of environmental laws was down more than 80 % under his watch! Trump's environmental record is such a toxic disaster it should be declared a Superfund site."
>
> — (Carol Browner, board chair of the League of Conservation Voters and a former EPA administrator) [53]

Trump is also quoted to have said that he wanted to nuke hurricanes and that windmills cause cancer? If we stopped chopping down all the forests and sucking all the water out of the environment for cattle, then we wouldn't have all this climate change in the first place! And nuking a hurricane could obviously, potentially cause cancer to humans by blowing radioactive isotopes everywhere and contaminating the environment. Besides poisoning people, the fallout from this activity would also kill millions of fish, birds and many other forms of sea life!

So, let's get the facts straight; to Trump, blowing up a nuclear bomb in the middle of a hurricane, so it can scatter radioactive contamination everywhere is safe, but windmills cause cancer?

Where's John McEnroe when we really need him? [54].

THE CURRENT PROBLEMS OUR ECONOMY FACES DUE TO NON-SUSTAINABLE PRACTICES:

The world today is at a critical epoch in history. If something is not done immediately, then the environment, along with the world economy, could fall into an abyss! Most of our non- sustainable economic woes are tied into an antediluvian energy infrastructure along with the cost and maintenance of an extremely profligate grain fed cattle complex.

Besides fossil fuel, many people are not aware that the production of livestock is wreaking havoc on all 6 continents of the globe, creating an environmental and economic nightmare! Livestock production is responsible for every ecological catastrophe we face; pandemics, pollution, endangered species, rain-forest destruction, wildfires, drought, famine, and antibiotic resistant strains of infectious diseases!

Here is a quote from sustainable author, Jeremy Rifkin;

"The devastating environmental, economic, and human toll of maintaining a worldwide cattle complex is little discussed in public policy circles. Most people are largely unaware of the wide-ranging effects cattle are having on the ecosystems of the planet and the fortunes of civilization. Yet cattle production and beef consumption now rank among the gravest threats to the future well-being of the earth and its human population." [55]

Due to all this wanton neglect of the environment, the world has been experiencing catastrophic natural disasters of greater and greater intensity; tsunamis, cyclones, tornados, wildfires, droughts and flash floods. Due to these global warming factors (with ice caps melting), the sea levels are expected to dramatically rise over the next 50 years, potentially causing trillions of dollars in property damage, contaminating freshwater resources and disenfranchising hundreds of millions of people who inhabit coastal regions around the world! Rainforests are being burnt down and destroyed primarily due to animal agriculture! Parabolic levels of man made methane, C02, ammonia and other effluvium emissions related to the cattle industry, have been documented by NASA. Methane traps 84 times more heat in the atmosphere than carbon!

Furthermore, there is herbicide contamination from too many feed crops along with acid rain, and many more anthropogenic related issues, all contingent upon livestock production [56]

According to Harvard scientists, in their keynote address entitled "Future of Life" it is scientifically explained how our carbon footprint is creating a cataclysmic impact on the earth's environment. It is their contention that we would need four planets to sustain life if all seven billion inhabitants of earth were to emulate the egregiously excessive lifestyle of the average American! An ecological footprint represents a person's toll on nature. The dynamics of this paradigm are calculated by using two key factors: (1) The amount of biological material consumed. (2) The amount of carbon waste created. This quantitative calculation determines your footprint. In essence, we must ecologically overcome this foreboding dichotomy if we someday wish to achieve total sustainability [57]

Left to its own device, greed will circumvent all efforts to eradicate climate change. In the book *Who Cooked the Planet,* author Paul Krugman of the *New York Times,* gives credence to this theory of demagoguery in science to cover up global warming for greed so that corporations can go on polluting by mitigating the facts! [58]

The advancement of green energy infrastructure has been dramatically impugned by avarice and collusion between big oil and multinational automobile manufacturing for more than a quarter of a century. (Refer to the movie *Who Killed the Electric Car.*) [59]

There is also the propaganda and demagoguery supporting the nuclear industry which inevitably could end up as a ...

"non-sustainable environmental and economic nightmare!"
(Refer to chapter "The Insidious Nature of Nuclear Power.")

Prolific political author Jeffery Sachs writes "We spent 8 years wasting our time and not putting any money into alternative energy sources, because we've spent a trillion dollars trying to defend the scarce resources in the Middle East." [60]

In 2007, financial columnist Hank Green published an article entitled "$7 Trillion: Cost of Not Acting to Prevent Climate Change." In this exposé, he emphatically pleads that we must spend 1% of the world's money now, or we will have to spend 20% later if we continue to procrastinate over the next forty years! [61]

Author Brian Dumaine in his book *The Plot to Save the Planet* explains "how visionary entrepreneurs and corporate titans are creating real solutions to global warming." [62]

PEACE AND PROSPERITY:

Society must make a commitment to a paradigm shift toward sustainability, and together with increased public awareness of sustainable practices, we can make a difference! Peace and prosperity will only be gained through a process of true sustainable ideals which must be established in order to forge a relationship into the future which meets the requirements of an ethical society. We will only have peace when we stop our dependence on foreign oil, and we will only have prosperity when we develop alternative energy in the United States, thus creating indigenous job growth!

Right now we are sitting at the forefront of a new revolution in renewable energy and sustainable transportation. If it wasn't for (as the Dalai Lama put it) *"all the greed, lies, and hypocrisy,"* then oil would already be at the brink of obsolescence. (Refer to "Veganomics: Economic Stimulus Plan.")

New electric vehicles are emerging on the scene that have hundreds of horsepower, can be driven hundreds of miles on a single charge, can be re-charged in as little as 15 minutes* [63] and only cost pennies per mile to drive. "WHAT ARE WE WAITING FOR?" As Gandhi said, *"If we want to make a change in the world, we must become part of that change."* Hydroelec-tric, wind, and solar technology is also proliferating. There are also huge tax credits available right now! It is these technologies that will make danger-ous, dirty, filthy, toxic, polluting, crude oil, coal, and nuclear energy a thing of the past!

"Nothing but blue skies from now on!"

Notes:

[1]   The Creature from Jekyll Island: "A Second Look at the Federal Reserve" by G. Edward Griffin, 2007.

[2]   www.amazon.com/The-Creature-Jekyll-Island-Federal/dp/. Pubs.usgs.gov/gip/prospect/goldgip.html.   [3]https://www.salon.com/2016/04/14/the_1_percent_are_the_real_villains_what_americans_ dont_understand_about_income_inequality_partner/

[4]   https://en.m.wikipedia.org/wiki/STOCK_Act

[5]   http://www.merriam-webster.com/autarky

[6]   https://quotepark.com/quotes/1109126-victor-hugo-no-army-can-withstand-the-strength- of-an-idea-whos/

[7]   Cronies: Oil, the Bushes, and the Rise of Texas, Americas Superstate books.google.com/books/about/Cronies.html?id=qFv0t61bvewC   [8]http://www.amazon.com/House-Bush-Saud-Rela-tionship-Dynasties/dp/ B000CC491W

[9]   www.amazon.com/See-No-Evil-Soldier-Terrorism/dp/140004684X

[10]   https://nypost.com/2013/12/15/inside-the-saudi-911-coverup/amp

[11]   https://en.m.wikipedia.org/wiki/Colin_Powell

[12]   Iraqi Defector Admits to Fabricating Story about WMD Claims," http://news.yahoo.com/s/yblog _thecut l ine/20110216/t s_yblog _thecut line/iraqi-defector- admits-lying-about-wmd-claims.

[13]   https://www.cbsnews.com/news/bin-laden-family-evacuate

[14]   http://www.sourcewatch.org/index.php?title=Blackwater: and The_Rise_of_the_World's_Most_Powerful_Mercenary_Army_(book)\. [15]http://www.fromthewilderness.com/.

[16]   http://en.wikipedia.org/wiki/saddam_Hussein_and_al--   Qaeda_link_allega-tions#cheney.27s_claims.

[17]   https://www.amazon.com/Crossing-Rubicon-Decline-American-Empire/dp/0865715408

[18]   https://en.m.wikipedia.org/wiki/The_Shock_Doctrine [19]

          http://blog.rapidsea.com/wp-content/uploads/2008/01/20080104-aclu-gitmoparadise.jpg. [20]https://theintercept.com/2018/12/01/the-ignored-legacy-of-george-h-w-bush-war-crimes-racism-and-obstruction-of-justice/

[21]   www.emkinstitute.org
         Speech Against the Invasion of Iraq

[22]   https://www.baltimoresun.com/news/chi-0303180333mar18-   story.html?output-
Type=amp

[23]   https://www.baltimoresun.com/news/chi-0303180333mar18-   story.html?output-
Type=amp

[24]   https://www.amazon.com/Age-Turbulence-Adventures-New-World/dp/0143114166

[25]   https://money.cnn.com/2006/01/04/news/companies/skilling/

[26]   https://www.thenation.com/article/archive/connect-enron-dots-bush/tnamp/

[27]   https://avalon.law.yale.edu/20th_century/eisenhower001.asp

[28]   Bill Maher show (7/24/09)

[29]   ibid.

[30]   ibid.

[31]   https://www.theatlantic.com/politics/archive/2017/09/giving-the-deep-state-more-   lee-
way-to-kill-with-drones/540777/

[32]   http://www.oilempire.us/reichstag-fire.html

[33]   https://www.nytimes.com/2001/09/24/us/a-nation-challenged-a-snapshot-gives-
bush-90- approval.html

[34]   https://en.wikipedia.org/wiki/Enabling_act

[35]   https://alumni.berkeley.edu/california-magazine/november-december-2006-life-after-
bush/law-john-yoos-war

[36]   https://www.history.com/topics/21st-century/patriot-act

[37]   https://amp.theguardian.com/world/2004/sep/25/usa.secondworldwar

[38]   Texas Executions: "GW Bush Has Defined Himself, Unforgettably, as Shallow and Callous," by
Anthony Lewis

[39]   John Paul Stevens, retired Supreme Court Justice. http://www.cbsnews.com/ sto-
ries/2010/11/23/60minutes/main7082572.shtml

[40]   http://www.ronsuskind.com/theonepercentdoctrine/.

[41]   https://www.justice.gov/archive/oip/foia_guide09/exemption6.pdf

[42]   https://www.cjr.org/public_editor/msnbc-public-editor-phil-donahue-and-the-art-of- re-
membering.php

[43]   https://en.wikipedia.org/wiki/Project_for_the_New_American_Century   [44]https://amp.
usatoday.com/amp/15440065

[45]   http://www.fromthewilderness.com/free/ww3/011805_simplify_case.shtml

[46]   http://grist.org/article/griscom-reagan/

[47]   http://www.getreligion.org///?P=587.

[48]   http://www.independent.co.uk/…/Jeff-sessions-attorney…

[49]   http://edifytrends.com/activists-wear-dinosaur-costumes…/

[50]   http://www.washingtonpost.com/…/is-trump-fighting-the…/

[51]   http://www.cnbc.com/…/chinese-billionaire-Jack-ma-says…

[52]   https://journals.plos.org/plosone/article?id=10.1371/journal.pone.0231929

[53]   https://amp.theguardian.com/environment/2020/oct/30/trump-agency-war-on- environ-
ment-former-epa-officials

[54]   https://gen.medium.com/amp/p/cd8348ed9770

[55]   Foundation on Economic Trends (FOET).org

[56]   https://news.un.org/en/story/2006/11/201222-rearing-cattle-produces-more-green-
house- gases-driving-cars-un-report-warns

[57]   Key Note Address: Harvard Scientists Warns Environmental Damage Irreversible" by Donald J. Johnston. http://www.oecdobserver.org/news/fullstory.php/aid/459

[58]   https://en.m.wikipedia.org/wiki/Paul_Krugman

[59]   http://www.sonyclassics.com/whokilledtheelectriccar/.   [60]https://www.marketwatch.com/amp/story/after-spending-trillions-of-dollars-on-wars-in- afghanistan-iraq-syria-and-libya-the-u-s-has-nothing-to-show-for-its-efforts-but-blood-in- the-sand-11629230262

[61]   http://www.ecogeek.org/content/view/318 [62]
         http://www.amazon.com/Plot -Save-planet-Visionary-Entrepreneurs/dp/0307406180

[63]   * Requires 440 volt outlet or charging duration will be significantly greater.

TheConsciousPlanet.org

*"They say that
I'm a dreamer
Well, I'm not
the only one!"*

— (John Lennon)

CHAPTER 6

VEGANOMICS

The World's Greenest and Most Compassionate Economic Stimulus Plan

Author Neil M. Pine, is a securities analyst and formally a Federally Registered Investment Advisor: SEC File # 801-41225

INTRODUCTION:

*E*very time you read this plan it makes more and more sense, not just for our national economy, but for the good of all humanity!

- This Plan should have unanimous bipartisan approval

- Will immediately stabilize the economy on Main St. and Wall St.

- Will not devalue the dollar, no new currency is created

- Will not create inflation, it merely fills an economic void stemming from a lack of buying power

- Uses existing Equity as a basis for capital

- Promotes business, industry, and job growth on both a micro and macro-economic scale. The largest corporations, to the smallest of business, will both see immediate stimulus.

- 80% of jobs are created by small businesses: Source: The Small Business Admin.

- Demonstrates an exponentially protracted and leveraged financial strategy.

- Will lay the groundwork for indigenous, long term, alternative energy infrastructure. The potential income and job growth created from these new technologies will lead to the obsolescence of fossil fuel, thus, breaking our dependence on foreign oil, and therefore displaying a new pride and peace of mind in America with the knowledge that:

1. It will bring back economic prosperity to America, taking the wealth and power away from Islamic extremists in the Middle East, and putting it to work in the U. S. E.g. North Dakota is the *Dubai* of wind power, California and Arizona are the *Iran* of solar power.

2. It will help to stop global warming for our future generations

3. It will stop our children from being killed in wars fought over foreign oil!

4. If we spent half as much money on developing safe, clean, alternative energy, as we have spent on all the wars fought over natural resources, mainly oil, and all the infrastructure needed to produce that oil, then we would probably have millions of additional jobs in the U.S. today, generating huge revenues with clean mega-watts of power!

The Dynamics of Proposed Economic Stimulus:

"Give a man some bread, and feed him for a day. Teach a man to sow and harvest, and feed him for a lifetime."

This popular adage helps to illustrate the concept of this financial paradigm. *(Assumes an average annual compounded growth rate of 10%)*

Step 1.) Financing:

The Federal Reserve, in conjunction with the U.S. banking system would play an intrinsic role in the implementation and promulgation of this system.

Step 2.) Equity:

Senator Bernie Sanders, recently stated that it's a hidden fact that 1% of the U.S. population owns more 90% of all the wealth in America. Therefore, it would be financially, ethically and environmentally prudent, to mandate legislation to reinvest a major portion of this wealth in alternative energy infrastructure. This is NOT socialism. The investment structure of this sustainable paradigm would not only be quite lucrative, but it is also totally necessary for the survival of humanity! There is NO Future being invested in non-sustainable technologies which cause pollution and promotes environmental destruction!

[a] One fifth (20%) of the asset value of the top wealthiest 1% of the U.S. population would be divested in alternative energy.

[b] One fifth, (20%) of the asset value of all U.S. based, non ADR oil companies would be divested,

Step 3.) Tax Free Investment Capital:

The top 1% of wealthiest people in the U.S., plus all U.S. based, non ADR oil corporations, would be required to divest 20% of all their assets into a tax free investment trust with a 30 year maturity note. This would be beyond what ever tax benefits each entity may currently be receiving. Due to reinvestment, it should be mandated that capital gains from such a divestiture should be voided. This would be similar to a (1031) exchange in real estate.

Operating under a tax-free environment, where investments can compound year after year, would not only produce a staggering proportion of growth, but it would also generate enormous secondary taxable revenue as a byproduct of increased industrial production and small business commerce. A 30 year tax free investment medium, (compounded annually), upon maturity, would yield more than 3 times that of investments which carry a 43% annual income tax structure for the wealthiest Americans.

Step 4.) Contingencies for Allocations of Tax Free Funds – Investing In Long Term Green Infrastructure and Indigenous Job Growth:

60% of this tax free investment capital must be committed to renewable energy and the job creation thereof. The other 40% should be allocated toward technical education and the creation of misc. small businesses which either cater to and/or compliment this new emerging green economy.

Step 5.) Immediate Short Term Economic Stimulus:

The Fed, in conjunction with the major banking institutions, would immediately allocate funds to the public, based on the original equity set up in the aforementioned tax free investments, in the form of alternative energy, green transportation (non-gas combustion vehicles), small business*, and first time home buyer tax credits*(*homes and businesses must be equipped with solar, wind or other renewable energy sources in order to qualify for stimulus).

This would immediately jumpstart the economy, thus creating a catalyst to support long term growth. The oil companies along with the top 1% of the wealthiest people in the U.S., at the end of the 30 year period, would pay back the Fed plus interest, at an adjustable prime rate for 30 years. This would be equivalent to a loan which is devoid of principal, interest, or tax structure for the life of the loan, but would only have one balloon payment (principal & interest), at the end of the 30 year period. Not only will these entrepreneurs' and titans of industry, help to bolster the economy, but they will also lay the ground work for future green infrastructure, which is desperately needed, yet still being neglected, by the current administration.

We wish to reiterate the powerful significance of this plan:

a. No capital gains taxes were paid on 20% wealth divestiture

b. No principle or interest payments for the life of a loan (Only 1 balloon payment after 30 year period)

c. Tax Credits: By giving away all the investment capital (based on the equity of the loan), then this would act as a major catalyst toward sustainability and the future of this planet. This plan is a major WIN, WIN for the investors and the future of humanity!

d. The long term compounded growth potential of these tax free investment vehicles, would easily pay for this short term stimulus.

Step 6.) Auxiliary Requirements for Oil Companies:
Building a Green Infrastructure

In order to qualify for the tax free benefits the oil companies must adhere to the following stipulations:

[a] Immediately divest 20% of assets and reinvest in alternative energy, (only hiring Americans in America).

[b] Make a commitment to be 50% invested in alternative energy by 2035

[c] Initiate an additional 30 cents per gallon tax. Every year thereafter, add 30 cents more on to this tax! Let the public know "Loud and Clear" that if you drive a vehicle that pollutes then you're going to have to pay Big Time, but if you choose to "Go With The Flow," and take advantage of tax credits and pollution free fuel economy, then you shall be greatly rewarded, and not just for your wallet, but for the environment as well! This money should go to help clean up the environment. This additional tax would further motivate consumers to take advantage of

the Economic Stimulus for alternative energy vehicles. If you are concerned about the driving range on EVs, then consider that a new technology is emerging, which claims to make the latest model Tesla run like a model T!

[d] To meet the 50% green target by 2030, the U.S. oil companies would be required to reduce their production of oil annually. Any noncompliance would be met with heavy fines.

Forcing the oil companies to make this paradigm shift is doing them a favor. Oil, without a doubt, will someday become obsolete and it already should have been if it weren't for all the "Greedy Gross Polluters" of the world! The cost of operating a gas combustion vehicle, as opposed to an electric vehicle, is more than 3 – 4 times greater.

Much of our economic woes are tied into an antediluvian energy infrastructure which has been impugned by greed, avarice and collusion, between Detroit (The tool and dye manufacturers for gas combustion engines), and the big oil conglomerates!

Step 7.) Auxiliary Trading Requirements for hedge funds, corporate traders, and large investors.

Over the last 30 years, the costs of almost all services have risen; from gardening to plumbing and even shining shoes. Everyone needs to make a living right? However, the cost of one service, that is contingent upon billions of dollars in revenue annually, has actually dropped by up to 100% during this same time period! This particular service, directly benefits the wealthy.

What could that be? A multi-billion dollar service that directly benefits wealthy people that is virtually free?

What is being referred to here, is that statistically speaking, the commission rates on publicly traded securities transactions (Due to the advent of internet technology), has declined by up to 100% in many cases, and especially for hedge funds in particular, who do the most trading. For example, TD Ameritrade, Charles Schwab, and many other major brokerage firms are all commission free for most large cap securities. In the 1970's, a trade with **Merrill Lynch** brokerage, may have cost over $100 for a just a few thousand-dollar trade, and thousands of dollars in commissions for large dollar trades.

Therefore, any investors, traders, or hedge funds should be required to pay a small tax commission. What seems very fair, would be a one tenth of one percent flat tax on trades under 100 thousand dollars and a one tenth of one percent or $300 flat fee (which ever is less), for any trades over one hundred thousand dollars, that would otherwise be commission free. However, even with this higher commission, the rate would still be roughly up to 98% lower than historical commission rates for these same trades. In 2015, Donald Trump, during his presidential campaign, announced a similar plan to tax hedge funds, which would generate billions of dollars in trading revenue.

Step 8.) Health Care Reform:

How the Government and the Food Industry is Deceiving the American Public and Making Them Sick and Fat!

Joe Biden has health care reform all backwards! We need to reform people's health, not their health care! Obviously, people must understand on a macro-economic level, that eating healthy has economic and physiological rewards. People who exhibit a healthy lifestyle are rewarded with longevity and a huge savings on medical bills. Isn't this a win- win situation for an individual and their insurance provider? The only losers are the Drug companies and the Medical industry, and for them may this author attempt to play the world's *smallest violin!* By analyzing at the root of our problems we can identify the proper motives and incentives to initiate this salubrious and sustainable paradigm.

The food industry (Which is heavily subsidized by the government), encourages people to eat unhealthily! You rarely see any ad dollars spent for vegetables, but for sodas, meats, dairy products, cookies, and fried salty snacks, they spend billions! We are bombarded by ads for these foods and see them everywhere we go; the bowling alley, car wash, super market, liquor or convenience stores, drive trough's (fast food), catering trucks, county fairs, theme parks, and family gatherings. Everything is loaded with GMOs (Roundup herbicide residue), sugars, saturated fats, cholesterol, hormones, MSG, antibiotics, etc. In today's modern western culture, it takes a lot of will power and an understanding of nutrition, in order to mitigate your exposure to all this insidious sustenance!

These government food policies further exacerbate gluttony and irresponsible eating habits. Has the government and the food industry conspired to make the American public sick and fat? In a special report by the late Peter Jennings, ABC Nightly News anchor and journalist, he exemplified this relationship between the government and major food conglomerates. The promulgators of these government subsidies never seem to factor in public health.

While examining the USDA food pyramid, the fats and sugars represent the smallest percentile at the top of the pyramid. The next stage, near the top of the pyramid, is represented by meat and dairy products, while the majority or bottom 2 sections are comprised of grains and produce. Of the total amount of money that supports American agriculture, less than 1% goes toward fruits and vegetables, including production and promotion. From a study conducted on the USDA from 1995-2002, the meat and dairy had received roughly 28 billion dollars in subsidies, while grains only made up about one third of that figure. During that same time period, sugars and fats received 3.8 billion in subsidies, while fruits were only allotted 170 million dollars, and vegetables got nothing. GMO Sugars and fats (Corn oil, soy bean oil, cotton seed oil, corn sweetener and beet sugar), received 20 times more subsidies than all fruits and vegetables combined! Monsanto/Bayer is currently being sued for multiple billions of dollars. More than any other corporation in history! Their pernicious weed killer, Roundup, has been directly linked to cancer. Therefore, our tax dollars go to subsidize products that cause cancer and make you sick!

> *"There is a disconnect between agriculture policy and health policy. The Government doesn't look at how agriculture policy can help improve public health. It's strictly about subsidies."*
>
> — (Tom Stenzel: United Fruits and Vegetables Association.)

Step 9. Sustainable Tax Credits:

Rather than to give trillions of dollars away to Drug, medical, and insurance companies; sustainable tax credits could be allocated to the public. With the allocation of sustainable tax credits along with public service announcements extolling the virtues of organic and sustainable foods, then we could someday see a dramatic reduction in the cost of health care, while simultaneously helping the environment and the local economy.

By educating the public as to the serious nature of environmental awareness combined with nutritional and health benefits, along with generous tax credits, then this type of program should prove to be quite lucrative. In essence, the government is paying us to stay healthy, while at the same time saving billions of dollars on health care and environmental cleanup. No matter who is running the country; we as consumers must make the final decision. We hold the power, through strength in numbers, to shape corporate America! The public must leverage their sheer numbers against the powers that be in order to send a message of sustainability. Sustainability is sensibility; it is the conscientious decision to change, but in doing so, we also evolve as more compassionate and ecologically minded human beings.

Benefits of Sustainable Tax Credits:

- Creates Demand for Sustainable Food

- Promotes Local and Regional Organic Farming

- Will Lower Cost of Health Care and Food

- Will Help to Stop Global Warming

- Promotes a more conscientious, compassionate & health conscious society

Step 10. Government Subsidized Bio-sustainable Supermarkets:

Government subsidized, Bio-sustainable supermarkets would create *home grown* job growth. Tax credits should be allocated toward the purchase of foods which are produced bio- sustainably. Bio-sustainable foods represent any form of sustenance which can be produced with as little harm to the environment as possible. E.g., locally or regionally grown organic produce and grains. Meat, even if organically produced, should still not qualify as sustainable.

By patronizing these establishments it helps put money back in the local economy, where we live, as well as reducing our carbon footprint. This in itself, should allow for the price of ecologically sound, organic produce to be reduced in price to more affordable levels, as mass (environmentally safe), organic agriculture production replaces (ecologically destructive), cattle ranching. Under current Federal guidelines, growing organic produce in many states constitutes land conservation, and therefore qualifies for tax credits. In addition, Legislation initiated by Senator, Sue Errington, has been active in the expansion of local, organic agriculture production and distribution.

These Sustainable supermarkets would represent a unique turning point in American sociology, in that for the first time in history, we will bridge the gap between less sophisticated grocery patrons (either due to naivety or indigence), with those of a more health conscious and/or greater social mobility, who can discern the benefits of an organic, vegan, macrobiotic diet. Buying organic produce should no longer be reserved for the affluent. Never before would underprivileged individuals be exposed to such a wonderful selection of organic grains and produce (free from irradiation, GMO's, pesticides, synthetic fertilizers, and chemical additives).

The end result of such an embourgeoisement (a shift in social consciousness of lower to working class citizens), would be a dramatic reduction in the cost of health care, while at the same time helping the environment and the local economy. This would represent a giant step toward a sustainable future.

Step 11. Food Stamp Revocation and Reform:

In 2001, a typical month represented 17.3 million U.S. welfare recipients at an annual cost of only 20 billion dollars. – Source: Michael Wiseman, (The Brooking Institute). As of 2006, a total of 5.7% of GDP, or 746 billion dollars had been earmarked toward promoting mobility. – Source: The Urban Institute. As of 2018, the total cost of poverty assistance had sky-rocketed to over a trillion dollars!

The Bio-sustainable food tax credits, given out by this stimulus plan, would be intended to supplement current welfare recipients, in addition to their current eligibility. However, due to the blatant misuse of Food Stamps, billions of dollars are lost or used for the wrong purposes. Under current regulations, if a person wishes, they can buy candy, soda, and Doritos', etc. with their Food Stamps. We see these big signs at fast food stores "EBT Cards Excepted." No wonder the United States, has the fattest poor people in the world! Today, the cost of health care, to welfare recipients is spiraling out of control!

By the revocation of current Food Stamp policy, and the substitution of bio-sustainable tax credits equivalent to each welfare recipient's eligibility, we could help to solve the unemployment problem by immediately creating jobs for local and regional farming interests, and at the same time, it would dramatically reduce the cost of health care by putting restrictions on un-wholesome food items.

Soda, candy, meat and dairy, should be banned to all Food Stamp re-demption. If these poor misguided individuals choose to kill themselves and their families through an egregiously unhealthy diet, then they better find the money from somebody else besides Uncle Sam!

Only non-processed, organic whole grain products and fresh organic produce, grown locally or regionally, would qualify for this new stimulus.

Some people may claim that this practice is unfair, and that they and their families are being deprived. And yes, it's true, they would be deprived: *De-prived of obesity, Deprived of cancer, Deprived of diabetes and Deprived of heart disease!*

With the propensity for an economic crisis looming, what sense does it make to give away food to the needy which may result in billions of dollars in increased health care as well as environmental destruction?

The sad truth is that most poor people don't seem to care about their health or the environment; they just care about their next meal! Bio-sustain-able tax credits would dramatically encourage all people to take better care of their health and the environment!

Ignorance is an inevitable byproduct of poverty. However, this ignorance is exacerbated and perpetuated through corporate subreption (A malicious con-cealment of the truth). The indigent, poor, or homeless, are a booming business to corporate America (junk foods, processed foods, and prescription drugs), es-pecially when being subsidized by Food Stamps or SSI!

The irony is that the biggest victims of this ignorance are the very people that Food Stamps were intended to protect: THE CHILDREN! By depriv-ing developing minds, bodies, and internal organs of proper natural suste-nance, serious health problems can develop, such as childhood obesity, type 2 diabetes, and learning disabilities which have plagued the inner cities of America for decades!

In addition, not only would bio-sustainable foods be good for the environ-ment, people's health, and the local economy, but it would also help to feed up to 5 times as many people! I wish to extrapolate upon this last statement due to the implied significance. E. g., someone who uses Food Stamps to buy ice-cream, soda, and 1 pound of hamburger. This would probably cost them approx. $10. If they purchased grains and produce of equal value, (beans, rice, corn, potatoes, greens, etc.), the vitamin, mineral, and fiber content could be up to 5 times greater, without the sugar, artificial colors, flavors, hormones, preservatives, pesticide residues, saturated fats, and cholesterol associated with these insalubrious foods of ignorance!

Also, non-perishable items like candy, coffee, or soda, can be more easily traded for drugs or alcohol! Believe me, it would be pretty tough to trade a bag of oatmeal and some broccoli spears for a hit of crack! Many desperate people, with substance abuse problems, may neglect their children by using their Food stamps to buy "non-perishable" items which can easily be traded for drugs!

It is unfortunate that most people feel in a truly free society, that there should be no restrictions against the use of horribly unhealthy foods or the non-sustainable practices which they represent, unless however, if it's with Taxpayers money! We should only subsidize compassionate and constructive utilization of food. We, as taxpayers, should not have to pay for peoples' self- destructive machinations, as well as the ecologically destructive and non-compassionate practices they patronize. And also, *don't forget about the children!*

Step 12. Tax Meat & Dairy Products:

The production of livestock and the consumption of animal products are some of the major causes of global warming, famine, desertification, pollution, and disease. There are so many tasty alternatives to animal products, free from hormones, cholesterol, and saturated fats. If people still wish to indulge in these insalubrious foods, then they should be required to pay an additional 20% sustainability Tax. This tax would go to clean up the environment and would also give people further motivation to *go vegan* and patronize bio-sustainable supermarkets.

This is one way to cut the pork from the economic stimulus!

Step 13. Geospatial Public Service Announcements:

Under the guise of the United Nations, the creation of international, digitally edited, public service announcements should be initiated. These (GPSA)'s could be displayed through all forms of multi- media, and would digitally translate into any language, giving a universal message of sustainability, warning people of the serious risks to their health and the planet if they don't change their dietary and lifestyle habits. Geospatial technology utilizes global software for civilian, business, and government applications, incorporating analytical methods with demographics.

Step 14. Compassionate Population Control:

We all know that over population is another non-sustainable factor threatening the survival of our planet. International sanctions should be initiated which mandate that each family may only have 1 child, unless of course, they are adopted. Incentives should be initiated to curb child bearing. Special tax credits should be awarded to families who sacrifice the bearing of their own children in order to adopt. However, if people really want to have more than 1 child then they should pay a Heavy Tax. This revenue would go into a Sustainability Superfund. In the future, Instead of buying a new Ferrari (which is bad for the environment), people will flaunt their wealth by paying huge penalties for having extra children. However, to be fair and equitable, free childbearing lottery tickets could be issued to underprivileged families who would qualify.

This system should be broken down as such:

1st Child: Free

2nd Child: $75,000

3rd Child: $150,000

4th Child: $250,000

Step 15. 420/ Compassionate Reform:

Drug laws are created to protect the public, especially when it comes to Health and Safety. However, marijuana seems to have an anomalous relationship, in regards to other substances, which are either illegal or restricted under the Health and Safety Code. Its innocuous nature and healing properties are synonymous among the terminally ill, which gives it a legitimate medical presence in today's society. These unfortunate people deserve to be treated with as much dignity and compassion and respect as possible.

A federal fund should be established in order to provide this medicine for people who would medically and economically qualify. A national data base of eligible people needs to be set up. Just because Big Pharma can't make a profit, then this should be NO excuse to deprive people of the best and safest medicine!

Medical Qualifications:

- *Terminal Illness*
- *Medically indigent or economically challenged medical status*

In Summary:

This plan boasts powerful economic benefits along with sustainable, philanthropic, and compassionate ideals. Under these guide lines it seems quite lucrative for the investors as well as the recipients of the stimulus. And in turn will bolster the economy, and free us from the Draconian foreign energy policies of the past. At the end of the 30 year period, investors would split up the profits left after the initial loan, plus interest, had been paid. You see, the banks are actually loaning their money to the investors of the stimulus plan, with the investors promise to pay it back with interest in one lump sum, after the 30 year period is up. These key investors would represent the heads of corporations and the titans of industry.

Certainly, their equity, based from a tax free investment medium would be solvent. Taking into consideration the tax benefits, job growth, and profits associated with such a system, combining this with the fact that the loan given is 100% backed up by real equity, and that the asset value of the loan recipients are at least 5 times greater than the loan amount itself, then this should spell out an overwhelming hypothesis of success, with almost a zero % chance of default! To sum it up, you may wish to refer to this plan as…

Green Capitalism on Steroids for the Good of Socialism.

The Conscious Planet

"The Earth Lies Polluted
Under its Inhabitants;
For They Have
Transgressed Laws,
Violated the Statutes,
Broken the Everlasting Covenant.

Therefore a Curse
Devours the Earth,
and its Inhabitants
Suffer For Their Guilt"

— Isaiah (24:5-6)

CHAPTER 7

POLLUTION
Livestock, Petrochemicals, Heavy Metals, Fluorides, Dioxins, and PCBs

Pollution usually consists of unnatural man-made toxins which are imbued throughout the environment in many different forms. These sources of pollutants can be disseminated by air, water or land.

However, not all pollution is necessarily created by man. Volcanoes for example, throughout history, have been linked to serious forms of air pollution and even climate change. The entire world's discharge of Volcanic pollution rivals that of fossil fuels and gas combustion vehicles. Besides ash and other incendiary properties, Volcanoes can produce enormous quantities of toxic green house gasses, such as Carbon Dioxide, Sulfur Dioxide, Hydrogen Sulfide, Carbon Monoxide, Hydrogen Chloride, and Hydrogen Fluoride.[1]

Red Tide: Another natural form of pollution which dramatically affects sea life is referred to as the Red Tide. The Red Tide is usually a naturally occurring toxin, characterized by the creation of toxic algae under the scientific name (Dinoflagellates). This type of toxic bloom produces (Satitoxin) and (Brevetoxin), which can wipe out entire fish populations. (See Endangered Fish chapter). Scientists have not yet determined the actual cause of Red Tide, though it has been concluded, that while some Red Tide outbreaks are completely natural, others have been directly linked to human activity.[2]

BASIC TYPES OF POLLUTION:

1. Air Pollution: When atmospheric substances and/or particles culminate in concentrations sufficient enough to jeopardize human health, then this constitutes air pollution.

   a.   Major Sources of Air Pollution:
- Transportation
- Power & Heat Generation

- Burning of Solid Waste
- Livestock Production
- Industrial Processing

b. <u>Examples of Air Pollution</u>:
- Exhaust Fumes from Gas Combustion Engines
- The Combustion of Coal
- Acid Rain
- Radiation (Refer to chapter "The Insidious Nature of Nuclear Power.")
- Cigarette Smoke
- Noise Pollution

2. **<u>Water Pollution</u>:** Water pollution is represented by the imbuement or discharge of chemical, biological, or physical materials which may impugn or degrade water quality, possibly affecting habitats and other contingent organisms.

a. <u>**Major Sources of Water Pollution**</u>:
- Pesticides & Other Petro-Chemical Compounds
- Heavy Metals and Other Toxic Metals
- Non-Degradable Solid Material
- Bio-Accumulative Substances

b. <u>**Examples of Water Pollution**</u>:
- Persistent Toxic Pollutants: Such as pesticides, heavy metals, petroleum, PCBs, Dioxins, and other chemical pollutants.
- Biological: Toxic medical, biological waste, and other manmade and natural substances.
- Physical: May be represented by the accumulation of innocuous dissolved or suspended solids.
- Agricultural Waste: Fertilizers, pesticides, herbicides, and fungicides, deposit their waste into oceans, rivers, streams, lakes, aquifers, wells, and riparian zones, also creating stagnant bodies of toxic quagmires.
- Industrial Production: The discharges of waste water used for industrial processing contain a variety of poisons, caustic acids, salts, alkalis, oils, and possibly harmful levels of bacteria.
- Mining: Acid water is the predominant by product in the production of gold and coal mining.

- Domestic Waste & Sewage Disposal:
- Radiation: (Refer to chapter The Insidious Nature of Nuclear Power)

3.   Soil Contamination: Land can suffer from various sources of pollution which may significantly affect bio-diversity and human activities.

a.   Examples of Land Pollution:
- Excessive Amounts of Manure from Livestock Production
- Chemicals
- Petroleum
- Herbicides (Weed Killers)
- Pesticides (Poisons which primarily kill insects and rodents)
- Fungicides (Inhibit the growth of molds)
- Radiation (Refer to chapter The Insidious Nature of Nuclear Power)

b.   Sources of Land Contamination:
- Poor Agricultural Practices (Livestock Production)
- Mineral Exploration
- Industrial Waste Dumping
- Oil Spills [3]

Manure Spills and Dead Zones:

There are approximately 1.5 billion cattle on the planet excluding other forms of livestock, each one being responsible for 40 pounds of manure per day. People say that manure is a necessary component for agriculture. While there may be some validity to this statement; there is also truth to the fact that approx. 1.2 trillion tons of manure is created annually on a worldwide basis. *"Either 'Bandini Mountain' is getting taller, or this planet is getting smaller!"* Statistically speaking, the production of livestock is responsible for twice the pollution of all industrial pollution combined! [4]

Due to a worldwide over production of livestock, combined with the enormous burden of industrial crop production, in order to feed farm animals, the oceans, and many fresh bodies of water around the world are being threatened from the runoff of fertilizer. This run off causes algae blooms to form in the ocean, where all oxygen is consumed, creating massive amounts of methane, killing fish and many other forms of sea life! The Methane gas traps 84 times more heat in the atmosphere than C02. There should be no doubt that global warming and climate change are directly related to this activity.

The disaster which befell Western Hungary, where caustic red sludge killed 9, injured 150, and caused property damage over a 15.6 square mile area, forcing evacuations, should be a wakeup call for the U.S. according to Alan Farnham of ABC News who states that "The Biggest U.S. Industrial Accident May be Waiting to Happen." [5]

Due to industrial farming practices, huge build ups of animal waste have accumulated in amounts of up to tens of millions of gallons to form dead zones. Statistically these dead zones double every decade. The imbuement and containment of all this waste has been an ongoing problem for decades, and is threatening the planet's freshwater resources, reserves, and wildlife habitats. In the U.S., many of these dead zones have been culminating over the years from Oregon to Chesapeake Bay. Here are a few significant examples.

North Carolina, 25 Million Gallon Hog Feces Spill:

Manure pollution has not only been responsible for global trouble, but also serious domestic problems as well. Historically, some of these industrial sewage dams have failed. One such manure reservoir failure, also referred to as manure lagoons, took place in 1995, in North Carolina, spilling approximately 22 million gallons of hog feces into the local watershed. This spill eventually infiltrated the New River, killing thousands of fish, before going on to pollute 364,000 acres of coastal wetlands. [6]

Louisiana, 8000 Sq Mile Manure Lagoon:

One of the largest manure spills in U.S. history took place in **Louisiana** and still encompasses 8000 sq. miles! This is equivalent to a car taking a 6-hour drive around the perimeter at an average speed of 60 MPH! Or if there were a bridge, it would take one and a half hours to drive across it at 60 MPH. However, *"Hold Your Breath"*; if it were possible to traverse across this dead zone, then you would not be able to breathe; all oxygen has been consumed! A person attempting this would die
of asphyxiation without proper artificial breathing apparatus.[7] [8]

Petro-Chemical Fertilizers (Nitrates):

Also, petroleum based fertilizers, (Phosphates), running off into our lakes, rivers, and oceans, are consuming all oxygen in their path. Petro-chemical fertilizers are poisoning water tables, wells, aquifers, lakes, rivers, streams, and riparian zones. This nitrogen compound creates an excessive enrichment of the water referred to as (Eutrophication), causing rapid algae growth which blocks out sunlight, thus creating a dramatic reduction of oxygen in water content.

Drinking water originating from arable areas has been found to contain high levels of nitrates and is considered harmful to newborn infants, as well as being a possible link to stomach cancer. [9]

Heavy Metals & Other Toxic Metals:

The culmination of heavy and other toxic metals has been a growing global environmental threat to humanity. The toxicity of metals has been directly linked to many medical conditions, ailments, and syndromes. Besides affecting internal organs and immune system function, Toxic Metals can also create Nero-toxic effects on the central nervous system. They scramble or short circuit signals given off by neurotransmitters to the brain.

Aluminum:

Even though Aluminum is not a heavy metal, it still contains seriously dangerous cumulative properties which are known to be responsible for a variety of ailments. One of the most serious of which is Alzheimer's disease, which statistically speaking, substantiates a 1400% greater aluminum content in the brain than what should be naturally occurring.

This toxic build up comes from food processing, containers, antacids, antiperspirants, baking powders and cheap confections, buffered aspirin, canned foods, city water supplies, cookware and utensils, cosmetics, foil for cooking, lipstick, air contamination from aluminum ore smelting plants, and processed cheeses. Besides Alzheimer's, aluminum toxicity has also been linked to Parkinson's, Dementia, and other neurological disorders.

Antimony:

This toxic heavy metal can be found in solder, sheet and pipe, bearings, metals, castings, and type metals. Also, in the form of Antimony Trioxide, is used in fire retardants for plastics, textiles, rubber, adhesive, pigments, and paper.

This obscure heavy metal is responsible for damage to the eyes, skin, respiratory system, circulatory system, and also causes damage to the blood, liver, kidneys, central nervous system, and reproductive organs.

Arsenic:

Arsenic can be found in a variety of sources. It's extremely poisonous and toxic properties can be found in meat and seafood products, chemical processing plants, cigarette smoke, unfiltered drinking water, fungicides, metal foundries, ore smelting plants, pesticides, polluted air, specialty glass products, weed killers, wood preservatives, and many other items.

This highly toxic, but colorless and odorless element can enter the body through the mouth, lungs, or skin. It can also affect the gastrointestinal system, also causing nervous disorders, respiratory illness, kidney damage, and cancers of the liver, skin, bladder, and lungs.

Cadmium:

This is another toxic heavy metal which also comes from many sources; some of which being air pollution, first and second hand cigarette smoke, batteries, ceramic glazes and enamels, tap and well water, food from cadmium-contaminated soil, fungicides, mines, paints, power and smelting plants, seafood and more.

The detrimental effects of this toxin occur mostly through inhalation or ingestion, and include danger to the lungs, kidneys, bones, and immune system. In some cases, cancers of the lungs and prostate may result as well as heart disease.

Essential Metal Over-Dose:

Our bodies require trace amounts of certain metals, but when taken in excess, these essential minerals can become highly toxic. Copper, Iron, and Zinc, are all prime examples of these potentially toxic elements. (See chapter 29: Vegan Macrobiotic Diest Section.)

Lead:

This toxic heavy metal used to be prevalent in gasoline and paint, but it still can be found in air pollution, ammunition, batteries, corrosive containers, contaminated soil, fertilizers, cosmetics, foods grown in contaminated soil, hair dyes, insecticides, lead glazed pottery, pesticides, solder, tobacco smoke, and water transported through lead pipes.

Lead attacks the central nervous system, kidneys, bones, blood, and heart. Lead can be particularly hazardous to pregnant women and newborn babies. It can affect fetal development, stunt growth, and cause behavioral defects and learning disabilities such as ADD.

Mercury:

This heavy metal toxin is synonymous with fish, but people can also be exposed to it through batteries, cosmetics, dental amalgam fillings, air pollution barometers, fungicides, insecticides, laxatives, paints, pesticides, shellfish, tap and well water, thermometers, thermostats, vaccines, and many other sources.

This highly dangerous substance has been linked with damage to the central nervous system, kidneys, and liver, along with causing Autism and Multiple Sclerosis. One third of all Mercury is naturally released into the environment, but man is responsible for the remaining two thirds of this toxic production.

The two major sources of mercury contamination stem from chlorine chemical plants and coal- fired power plants. And along with auto-scrap recycling, they generate well over 100 tons of mercury each and every year. The main sources of exposure to Americans are dental fillings, tuna, and other fish, and vaccines. (Refer to chapter "Shots of Ignorance") [10]

Petro-Chemical Pollution:

Sources of Petro-Chemical Pollution include Chemical Plants, Power Plants, Oil Refineries along with other Refineries, Natural Gas Processing, and Bio-Chemical Plants. Also Water and Waste Treatment utilize Petro-Chemicals

Petroleum: Hazardous Spills and Side Effects

The hazardous side effects of oil exploration have become apparent with the Exxon Valdez, and the B.P. (British Petroleum) oil spills. Oil spills are comprised of the discharge of liquid petroleum hydrocarbons into the environment, directly related to human activity. This termi-

nology is most associated with marine oil spills, where oil contaminates sea water. The Exxon Valdez was responsible for only releasing 11 million gallons of oil as opposed to a whopping record of 18-28 million barrels for the B.P. disaster, by far surpassing Exxon's old record of environmental negligence!

Ecological Effects: Birds are particularly at risk when exposed to oil spills. *"Less than 1% of sickened birds survive, even after cleaning."*- Wikipedia. Marine mammals are also seriously affected, their fur becoming matted with oil, seriously affecting their body's insulation and immune function. (See Endangered Species Section) [11]

PCBs:

PCBs, technically referred to as (Polychlorinated Biphenyl), are a class of organic chemical compounds which are utilized in many industrial applications, such as (Dielectric Fluids) for transformers, capacitors, and coolants.

In 1979, the U.S. Congress in confluence with the Stockholm Convention on Persistent Organic Pollutants, banned the production of these chemical compounds. Research studies have indicated that physical effects of PCB exposure is similar to that of Dioxin which are associated with (Endocrine) disruption and (Neurotoxicity).

PCBs display (Mutagenic) and toxic effects to the body through the interference with the production of hormones. This interference with hormone production has been linked to many forms of cancer, mainly affecting women. This is primarily due to these chemicals' estrogenic blocking qualities which increase the risk of breast, uterine, and cervical cancers. PCBs can also lead to developmental problems in both male and females, which includes mental, sexual, and skeletal development. [12]

Dioxins:

Dioxins represent a group of chemical compounds that are regarded as Persistent Environmental Pollutants. This chemical substance culminates in the food chain and deposits itself in the adipose tissue of animals. Over 90% of all dioxin exposure is through the food chain. Meat, dairy, and various forms of fish, especially shellfish and bottom dwelling fish, have the greatest risk to exposure from this chemical contamination. This chemical is also known to interfere with the production of hormones and is also linked to various forms of cancer.

Due to this ubiquitous exposure of **Dioxins** throughout the environment, efforts need to be initiated to reduce background exposure.

Sources of Exposure:
Uncontrolled waste in incinerators (Solid waste and hospital waste), due to incomplete incineration. [13]

<u>Sulfur Dioxide:</u>
Sulfur Dioxide, representing the chemical compound SO; it can be naturally released by volcanic activity, but most of its negative effects have been contingent upon man. Sulfur Dioxide is a strong-smelling poisonous gas utilized by industry and is also a major factor in the creation of air pollution. Due to this element entering the earth's atmosphere, it creates elevated levels of hydrogen ions. This starts a process referred to as "Wet Deposition," when the water molecules become imbued by Sulfur Dioxide and Nitrogen Oxides, thus producing what is referred to as Acid Rain.[14]

Historically, Acid Rain has had extremely detrimental effects on the environment. Since the 70's, governments worldwide, have made serious efforts to curb the release of this industrial air pollutant. However, Acid rain has still been an ongoing problem in developing nations such as China, India, and Russia, where coal plants are prevalent.

Acid Rain demonstrates negative impacts on forests, fresh waters, and soils. It kills insects and various forms of aquatic life while causing damage to buildings and impacting human health. [15]

FLUORIDE: INDUSTRIAL WASTE SCAM:

During the 1960's and 70's, industrial pollution wreaked havoc on the Mohawk Indians of the New York-Canadian St. Regis Reservation.

Cows crawling on their bellies like giant snails, so crippled by bone disease that they could no longer stand up. Their teeth, crumbled down to the nerves, no longer enabling them to properly Masticate (Chew their cud), in order to assimilate nutrition from high fiber forage. These cattle inevitably starved to death.

This industrial pollution devastated the Mohawk people. Besides killing their cattle, crops and trees withered, the birds and the bees fled, and fish from the nearby St. Lawrence River had contracted physical ulcerations and spinal deformities. In addition, the Mohawk children also had suffered from damage to their bones and teeth.

ALCOA Aluminum, along with Reynolds Metals Corp., was subsequently sued by the Mohawk people for 150 million dollars for loss of property and health concerns. But after 5 years, legal expenses had bankrupted the tribe. Physically and psychologically defeated, they settled for a mere $650,000.00. Is this how we pay homage to aboriginal Americans? This is typical environmental courtroom demagoguery, making evil polluting corporations look like saints while the victims of their sinister greed driven machinations, are humiliated and belittled into obscurity! When quoted, human rights lawyer, Robert Pritchard stated *"What judge wants to go down in history as being the judge who approved the annihilation of the Indians by Fluoride emissions?"*

So then why is it that for nearly 50 years the U.S. government and the media have extolled the virtues of Fluoride? Toiletry manufacturers began putting it in their toothpaste, and city municipalities started tainting their drinking water with it.

Smear campaigns and demagoguery was initiated against any group or individuals who questioned Fluoride's safety. People, who opposed Fluoride, then and even today, were dismissed by the government and the media as quacks or lunatics. What the government didn't tell you is that fluoride is actually a toxic industrial pollutant. It is one of the largest and oldest sources of toxic waste. Due to this serious environmental damage, most modern fluoride pollution plants have only been forced to reduce emissions, but not eliminate them. Even today on this Mohawk territory, cows still only live one half their normal lifespan. The Environmental Protection Agency, (EPA), has estimated that at least 155,000 tons of fluoride is released into the atmosphere every year through industrial plants. Another source of fluoride contamination is in the water. Roughly 500,000 tons a year of toxic fluoride, are released into rivers, lakes, and oceans.

People living nearby or working in heavy fluoride emitting industrial plants, have received high dosages. Fluoride is a non-biodegradable substance. It culminates in the environment, working its way into the food chain, poisoning people's bodies, and settling in the bones and teeth.

Industrial polluters have gone through inexorable legal measures to ensure a mitigation of their environmental responsibilities. The production of

Steel, Iron, Aluminum, Copper, Lead, and Zinc, all generate serious levels of fluoride emissions. Fluoride emissions are also a contingent factor in the production of (Phosphates), (which are critical to the manufacturing of non-organic fertilizers), Plastics, Gasoline, Brick, Cement, Glass, Ceramics, and all Clay products.

Uranium processing, electrical power generation and coal combustion are also gross polluters of Fluoride. Both industry and government have carried a clandestine motive for claiming that contaminating our toothpaste and drinking water with Fluoride is beneficial to humans.

However, this claim has not been met without opposition, due to fluoride being the world's most dangerous industrial pollutants. It is by far, the most toxic to humans, animals, and vegetation.

And pound per pound, is considered one of the most toxic substances known to man. Hitler also put fluoride in concentration camp victims' drinking water, but it was not to help them with their smile!

According to the United States Department of Agriculture (USDA), *"Airborne fluorides have caused more worldwide damage to domestic animals than any other air pollutant"* In regards to vegetation; even in reports dated back to 1901, research found that *"Fluoride compounds are much more toxic than other compounds that are of significance in the industrial smoke problem."* Not only has our government dismissed information which was provided as early as the 1930's, warning people about the dangers of Fluoride toxicity, but it has actually promoted a malicious intent with the intentional poisoning of the nation's drinking water with recycled industrial waste!

Fluoride started to gain notoriety in 1933 when the world's first major air pollution disaster

Source: www.NaturalNews.com

hit Belgium's Meuse Valley. Several thousand people became violently ill, and

60 died. At that time, Kaj Roholm, the world's leading authority regarding fluoride, blamed fluoride for this disaster.

At that time, health scientists began regarding fluoride as a deadly and toxic poison. This negative environmental awareness of fluoride toxicity could have been potentially hazardous to the profit margins of many industries who were gross polluters of this toxin. In 1933, Lloyd DeEds, Senior Toxicologist with the (USDA), wrote *"Only recently, that is, within the last 10 years, has the serious nature of fluoride toxicity been realized"*

Government and industry leaders met to discuss an insidious mass subreption against humanity! The future of military and U.S. industrial expansion, and the huge profits it represented, was contingent upon contaminating the environment with millions of tons of toxic fluoride industrial waste. Two major military industries positioned for industrial expansion were the Aluminum and the Fluorocarbon Industries. Fluorides were major by-products of these industries.

In 1938, the proliferation of infrastructure supporting these industries was critically important to the upcoming world war. Experts warned industry leaders that rapid expansion of industrial production would create unprecedented harm to the environment, animals, and human health, and that the risk of lawsuits could be financially devastating to many companies.

With the onslaught of WWII, came the birth of the "Military Industrial Complex," with all its concomitant public demagoguery campaigns. Industrialists and military defense contractors created a "Federal Blitz Campaign," to propagandize the public into believing that fluoride was actually good for them. In 1939, a major national public announcement by a scientist working for ALCOA, Gerald J. Cox, stated *"The present trend towards complete removal of fluoride from water and food may need some reversal"*

Mysteriously, many new research reports started to emerge on the scene extolling the virtues of fluoride. One such bogus research center, funded by ALCOA, Reynolds, and other gross emitters of fluoride, was the Kettering Labs. A book written by E.J. Largent, a Kettering scientist, and consultant for Reynolds Metals, was written in part to *"Aid industry in lawsuits arising from fluoride damage."* This literature became a major international reference work. The Big Lie being promoted was that fluoride was beneficial to human health and children's teeth, in low dosages. It was proposed to add fluoride to the entire nation's drinking water. This

(Fluoridation) on a national scale, would require an additional hundreds of thousands of tons of fluoride to be added into our country's drinking water. Major industries, especially ALCOA, strongly supported government water fluoridation projects. If they could convince the public that fluoride was *"a health-enhancing substance which should be added to the environment for children's sake"*, then any groups or individuals opposing this view would be labeled as quacks or lunatics.

ALCOA Circumvents Accountability:

Overnight, fluoride became the *"Protected Pollutant,"* as writer, Elise Jerard coined it. During the roaring 20's, the U.S. Health Services fell under the jurisdiction of Treasury Secretary, Andrew J. Mellon, who was also founder and a major shareholder of ALCOA. In 1931, Mellon stepped down from this position, but not before installing a Public Health Service dentist named H. Trendley Dean to do PR work for ALCOA. Under Dean, research was conducted to determine just how much fluoride people could tolerate without obvious damage to their teeth. Dean found that people's teeth, in towns that had higher concentrations of fluoride in well water, suffered from a greater propensity of discoloration and erosion, but also statistically reported fewer cavities.

Thereafter, a study, at the Mellon Institute, (ALCOA's) Pittsburgh industrial research lab, concluded that fluoride reduced cavities in lab rats. ALCOA sponsored biochemist, Gerald Cox, took credit for this report. So therefore, in 1939, the first nationwide public proposal to fluoridate America's drinking water, did not come from a dentist or doctor, but rather an industry scientist working for a company which had more to lose than any other company regarding litigation from fluoride contamination!

Hundreds of post war fluoride damage lawsuits were filed around the country against the producers of Steel, Aluminum, Iron, Copper, Phosphates, and other Chemicals. One such case involved an Oregon couple, *Paul M. And Vera Martin v. Reynolds Metals* (1955). The court ruled that the couple had sustained *"serious injury to their livers, kidneys, and digestive functions"* from eating *"farm produce contaminated by [Fluoride] fumes"* from an adjacent Reynolds Aluminum plant. However, after this court precedent decision, ALCOA, and six other metal and chemical companies, all joined in with Reynolds Metals, officially calling themselves, "Friends of the court," and overturned that decision. A Reynolds attorney, when interviewed for the local paper, *"contended that if allowed to stand, the verdict would become a ruling case, making every aluminum and chemical company liable to damage claims simply by operating* [emphasis added]." But despite corporate influences, the Martin's were able to reverse court rulings again. So finally, to avoid historical court precedence, the Reynolds Corporation bought the Martin's ranch for an exorbitant sum of money.

Postwar damages to forests, livestock, and smog stricken urban residences, were prolific. But little or no public attention was ever given. All this bad publicity had been diverted away by Fluoride's Sparkling New Image. Shortly before the end of the war in 1945, the federal government selected two cities in Michigan for a comparison study to determine if fluoride could actually, safely reduce cavities in children. Thereafter toxic fluoride was pumped into the drinking water of Grand Rapids.

And in 1946, Regardless of the fact that the official 15-year experiment had just started, six more U.S. cities started fluoridating their water. This insidious movement began to gain momentum. And at this time an extremely experienced attorney working for ALCOA, Oscar R. Ewing, had just been appointed head counsel; his fee's reaching an un-precedented $750,000 a year. He had previously been instrumental in government negotiations for ALCOA's war time plants. In 1947, Ewing was appointed head of the Federal Security Agency (later referred to as HEW), which also gave him jurisdiction over Public Health Services (PHS).

Under his tainted regime, the (PHS), by 1950, had initiated fluoridation in 87 additional cities, including the control city used for the original 1946 experiment, thus extirpating the most important objective, (To test safety & benefit) of the experiment before it was even one third under way!

SIGMUND FREUD'S NEPHEW: EDWARD L. BERNAYS

The (PHS)'s excuse for their unscientific haste was "popular demand." And at that time Oscar Ewing had hired one of the most influential public relations gurus in history; Edward L. Bernays, Sigmund Freud's nephew, who utilized his uncle's theories towards the media and government propaganda. He was known as "The father of public relations," and the *Washington Post* had recently called him "The Original Spin Doctor."

This fluoride campaign represented one of his most stellar and triumphant successes.

In his book, *Propaganda*, written in 1928, Bernays goes on to explain *"the structure of the mechanism which controls the public mind, and how it is manipulated by the special pleader [i.e., public relations counsel] who seeks to create public acceptance for a particular idea or commodity. Those who manipulate this unseen mechanism of society constitute an invisible government which is a true ruling power of our country ... our minds are molded, our tastes formed, our ideas suggested, largely by men we have never heard of..."*

Bernays writes *"If you can influence the [group] leaders, either with or without their conscious cooperation, you automatically influence the group which they sway."*

EINSTEIN'S NEPHEW: DR. E.H. BRONNER:

While prolific intellectuals like Freud's nephew were paid exorbitant amounts of money to extol the virtues of fluoride, another equally brilliant man's nephew, Albert Einstein's, was a strong anti-proponent of fluoride, and received no or very little monetary compensation in doing so.

Dr. E. H. Bronner, nephew of Albert Einstein, was a scientist and research chemist. Here he makes a statement regarding fluoride's safety:

> *"Even in very small quantities, sodium fluoride is a deadly poison to which no effective antidote has been found. Every exterminator knows that it is the most effective rat-killer. Sodium Fluoride is entirely different from organic calcium-fluoro-phosphate needed by our bodies and provided by nature, in God's great providence and love, to build and strengthen our bones and our teeth. This organic calcium-fluoro-phosphate, derived from proper foods, is an edible organic salt, insoluble in water and assimilable by the human body; whereas the non- organic sodium fluoride used in fluoridating water is instant poison to the body and fully water soluble. The body refuses to assimilate it." "That any so-called 'Doctors' would persuade a civilized nation to add voluntarily a deadly poison to its drinking water systems is unbelievable. It is the height of criminal insanity!"*
>
> — (E.H.Bronner) [16]

"There's a sucker born every minute"

— (WC Fields)

This popular cliché describes how a chemical used as rat and bug poison, magically overnight, becomes safe and good for children, according to our trustworthy and benevolent Big Brother. Wow, and it's Free! Aren't we lucky that they don't charge us to poison us? Oh, but if there were only government transparency in taxation, you would find out that they actually do!

Into the 60's and 70's, a dichotomy of opposition to fluoride in drinking water came from the right and the left wing. The John Birch Society and the Ku Klux Klan, expostulated that fluoride was a communist plot to dumb down America. Also, educated scientists, environmentalists, Left wing, and new age thinkers were also concerned about this deluge of (PHS) fluoridation. Because the fluoride movement was sponsored by such credible organizations as the American Dental Association and the Metropolitan Water District, any opposition had been met with smear tactics and demagoguery.

In 1950 the (PHS) in conjunction with industry leaders of dentistry, medicine, and many other fields, gave fluoride their official endorsement. This unbelievably insane and sinister transformation was complete. The public was brainwashed, fluoride only kills bugs and rodents, but is good for children and humans? Since this time, two thirds of our nations' reservoirs have been contaminated by this deadly toxin. Roughly 143,000 tons of fluoride is pumped into our nation's water supply annually, and our government continues to maintain its stance on fluoridation practices, even though *"The same concentrations added to human drinking water for cavity protection can be fatal to fish"*

Regardless of this malice aforethought by our government against the American people, it was only a matter of time before more new scientific evidence surfaced. In 1983 this evidence presented itself, and in response, a government appointed commission was formed. These committees are typically comprised of veteran fluoride defenders, with never an opponent in sight. The perfunctory mission of these committees is always to relegate and dismiss new evidence, therefore reestablish the status quo. However, due to pressure in this case, the (PHS) was forced to convene a panel of "World Class Experts" in order to conduct Pro forma review of safety data on fluoride in drinking water. According to the panel transcript by its private members, reveals that, very little evidence exists which substantiates fluoride safety claims.

Caution, especially to children was recommended by this panel. However, Jay Shapiro, working on behalf of the National Institute of Health, (NIH), understood that the recommendations of this panel conflicted with government fluoride policy. in an official memo drafted by **Shapiro** he remarked *"[B]ecause the report deals with sensitive political issues which may or may not be acceptable to the PHS [Public Health Service], it runs the risk of being modified at a higher level."*

With all findings taken into consideration, Surgeon General Everett Koop's office, one month later, reported the panel's recommendations. Mysteriously, all their most serious recommendations had been discarded, without mention, and apparently without even consulting with its members. Another member of the panel, Daniel Grossman, wrote *"When contacted, members of the panel assembled by the (PHS) expressed surprise at their report's conclusions: They never received copies of the final---altered--- version. A scientist from the Environmental Protection Agency (EPA), Edward Ohanian, who had observed and documented the panel's findings and deliberations recalled being 'baffled' when the agency received its report."*

All recommendations for low dose **fluoride** warnings were thrown out. Substituting for their research was a blanket statement: *"There exists no directly applicable scientific documentation of adverse medical effects at levels of Fluoride below 8ppm [parts per million]."*

This statement was in direct conflict with the committee's final draft, totally contradicting the panel's recommendations, which clearly stated that ...

> *"the Fluoride content of drinking water should be no greater than 1.4 -2.4 ppm for children up to and include age 9 because of lack of information regarding fluoride effect on the skeleton in children (to age 9), and potential cardio toxic effects [heart damage]."*

Quoting from a transcript of the panels meeting: *"Dr. Wallach: You would have to have rocks in your head, in my opinion to allow your child much more that 2ppm."* In response to this statement, Dr Wallach replied *"I think we can all agree on that."* Yet not long after this meeting, in 1985, based on the altered report issued by Surgeon General Koop, the EPA increased the level of fluoride in drinking water from 2 to 4 ppm for everyone, including children!

Recent reports published in the Journal of the American Medical Association (JAMA), found increased risk of bone fracture, and other health risks due to water fluoridation. According to the National Research Council, the U.S. has the highest rate of hip fracture in the world.

Evidence from government animal research studies, seem to indicate that carcinogenic effects of fluoride contamination have been in existence since the 40's. But in 1956, the government dismissed all such claims. According to research conducted in 1975 by John Yiamouyiannis, a biochemist and controversial fluoride opponent, along with Dean Burk, retired National Cancer Institute (NCI) official, reported a 5-10% increase in total cancer rates in U.S. cities which have fluoridated their water supplies. This report triggered congressional hearings in 1977, and to their chagrin, no governments tests for cancer for fluoride had ever been issued! Later that year Congress ordered the NCI to begin testing for carcinogenic effects of fluoride contamination. In conclusion of this study, 12 years later, the NCI found "equivocal evidence" that fluoride caused bone cancer in male rats. The NCI immediately began to focus on nationwide cancer trends related to fluoride. Direct evidence from international sources pointed out a direct correlation of higher rates of bone and joint cancers in males from fluoridated countries.

In another study conducted by the (NCI), on males and females under 20, in several counties of Washington and Iowa where the water had been fluoridated, the rates of bone and joint cancers increased by 47% from 1973-80 to 1981-87. However, in the non-fluoridated sections of these states, the rates of these aforementioned cancers declined by 34%! For bone cancer (Osteosarcomas), the rate increased by 70%, while in the non-fluoridated counties of these two states, bone cancer decreased by 4%. Another report issued in 1992, by EPA senior services advisor, William Marcus, points out a 600% increase in the rate of bone cancer among young males in New Jersey living in fluoridated communities. In a statement issued by Marcus, he said, *"In my opinion, fluoride is a carcinogen by any standard we use. I believe the EPA should act immediately to protect the public, not just on cancer data, but on evidence of bone fractures, arthritis, mutagenicity and other effects."*

Once again, a new **(PHS)** commission was formed by highly respected **fluoridation** proponent, Frank E. Young. In this report, he concludes that *"its yearlong investigation has found no evidence establishing an association between Fluoride and cancer in humans."* In this report Young also stated that *"The U.S. Public Health Service should continue to support optimal fluoridation of drinking water."* [17]

Notes:

[1]   http://www.thenakedscientists.com/forum/index.php?topic=15196

[2]   http://www.upi.com/Science_News/2009/01/07/Red-tide-linked-to-nutrient-pollution/ UPI- 60321231350271/

[3]   https://www.worldatlas.com/articles/how-many-types-of-pollution-are-there.html[4] https://www.newsweek.com/2015/12/18/two-numbers-animal-manure-growing-headache-america- 402205.html?amp=1

[5]   http://abcnews.go.com/Business/PersonalFinance/industrial-accident-waiting-   happen/ story?id=11

[6]   http://www.aphis.usda.gov/animal_health/emergingissues/downloads/1pigs.pdf

[7]   http://thegreenestlittlehouseintexas.org/Agricultural_Water_Use.html

[8]   Pools of Animal Waste!… | Jetsetting Magazine
           http://blog.jetsettingmagazine.com/environment/pools-of-animal-waste

[9]   http://www.talktalk.co.uk/reference/encyclopaedia/hutchinson/m0035393.html

[10]   Amazon.com: Genetic Heavy Metal Toxicity: Explaining SIDS, Autism, Tourette's, Alzheimer's and Other Epidemics (9780595480562):

           http://www.amazon.com/Genetic-Heavy-Metal-Toxicity-Explaining/ dp/059548056X/ref=sr_1_2?s=books&ie=UTF8&qid=1308122347&sr=1- 2#reader_059548056X

[11]   http://en.wikipedia.org/wiki/Oil_spill

[12]   http://en.wikipedia.org/wiki/Polychlorinated_biphenyl

[13]   www.who.int/mediacentre/factssheets/fs225/en/

[14]   https://www.epa.gov/acidrain/what-acid-rain

[15]   http://www.macmillandictionary.com/dictionary/british/sulphur-dioxide

[16]   http://www.nowpublic.com/health/fluoride-not-what-your-think-its-very-bad   [17]   http://www.maebrussell.com/Fluoride/Fluoride%20%20Commie%20Plot%20or%20Capitalist%2 0Ploy.html

The Conscious Planet

"Pollution, Pollution,
you can brush your teeth
with the latest toothpaste,
and then rinse your mouth
with industrial waste
Cha, cha, cha!"

– (Tom Lehrer)

CHAPTER 8

CATTLE AND EGREGIOUS GREENHOUSE GASSES

Since the beginning of the industrial revolution in 1750, the rate of CO2 created in the atmosphere has been constantly rising. However, by the early 1970s, the yield curve of CO2 production had reached parabolic levels. And with burgeoning world economies, such as in China and India, this yield curve continues to grow exponentially! The failure to stop global warming will inevitably result in dramatic increases in extinction rates as well as hundreds of millions of displaced and starving refugees! Also see chapter 8 on famine "Overfed Cattle = Starving People.") [1] & [2]

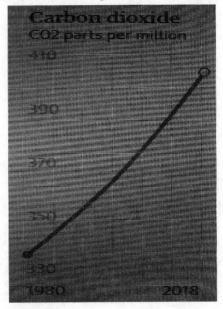

<u>1980</u>   <u>2018</u>   <u>Carbon Dioxide</u>

In an article published by the United Nations on November 29th, 2018, entitled: "Tackling the World's Most Urgent Problem: Meat – UN Environmental Program"

It has been scientifically documented that global warming gasses from livestock production far exceeds that of all gas combustion vehicles. All life on earth is contingent upon the delicate balance of atmospheric mechanisms which we all take for granted. This fragile system of physics and chemistry is being threatened. Livestock are creating egregious anthropogenic effects over the earth's biosphere. Climate change from livestock production is causing cataclysmic shifts in weather patterns and ocean currents, leading to melting polar ice caps and glaciers, which create rising water temperatures and sea levels, while simultaneously threatening human and wildlife habitats. Producing beef is 34 times more polluting to our climate than beans or lentils. [3]

9/17/2021, Biden announces a global plan to reduce planet-warming methane emissions! The European Union and the Biden Administration have launched a joint effort to reduce methane gas in the atmosphere, 30% by 2030! This announcement was made in advance of a critical climate conference by the U.N. in Glasgow in November of that year. [4]

OCEAN SURFACE TEMPERATURE CHANGE

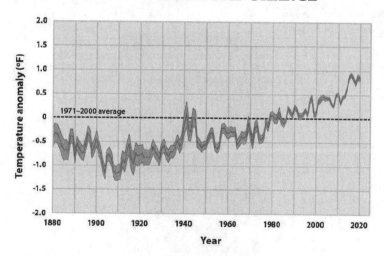

Arctic Melt Down

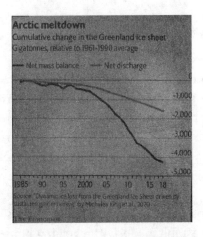

As we know, the Trump Administration had blatantly denied the existence of manmade climate change. Also, evidence had surfaced that under the former Bush administration, reports by the National Oceanic and Atmospheric Administration warning of a strong link between powerful hurricanes and climate change were routinely blocked. (Refer to chapter 4 "Peace and Prosperity") Further research published from scientists at the National Center for Atmospheric Research in Boulder, Colorado, also concluded that human activities are responsible for severe climate change resulting in rising seas and cataclysmic weather patterns. (See chapter 8 "Dust Drought and Desertification) [5]

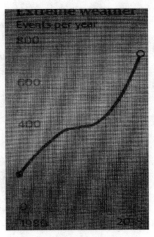

1980   2018   **Extreme Weather**

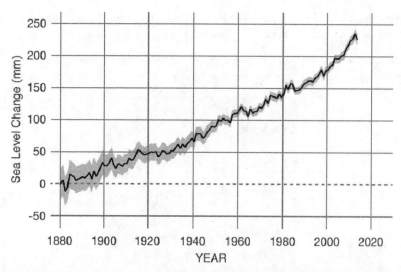

1880 2020 Sea Level Change Methane and Other Global Warming Gasses:

Dead zones in the ocean are contingent upon massive amounts of fertilizer runoff and are creating enormous amounts of methane gas in the atmosphere. Statistically speaking, the livestock complex is responsible for 37% of all methane gas emissions. Methane gas can potentially trap up to <u>84 times</u> more heat in atmosphere than CO2. Livestock are responsible for 18%*-51%** of all greenhouse gas emissions measured in CO2 equivalent. The production of cattle and other ruminants also emits 65% of all nitrous oxide released into the atmosphere. This is equivalent to up to 266 times the global warming damage potential of CO2. The majority of these emissions are due to manure. Also 64% of ammonia is attributed to animal waste which is also a contingent factor in the creation of acid rain. Manure management, just by itself, is attributed with 14% of all agricultural greenhouse gas emissions in the U.S.! [6]* & [7]**

> *Livestock are one of the most significant contributors to today's most serious environmental problems. Urgent action is required to remedy the situation.*
>
> — (Henning Seinfeld, official for UN Food and Agriculture Organization, or FAO. (See chapters "Cattle and Pollution" and "Dust Drought and Desertification.") Also refer to "Livestock's Long Shadow: Environmental Issues and Options." [8]

The world wants to emulate the United States with multinational fast-food establishments infesting all corners of the globe. Statistically, with this Westernized convenience, the sale of meat and dairy products has been constantly rising with demand projected to double by 2050 [9]

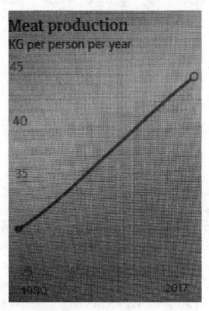

| 1980 | 2017 | Meat Production |

There are also serious issues with cow, and even termite flatulence, which also affect meteorological conditions. The real concern over damage to the earth's atmosphere is no longer from ozone layer depletion created by chlorofluorocarbons (CFCs), but rather from the greenhouse effect due to livestock production. In an article published in the *Washington Post* entitled "Feed, Animal Flatulence, and Atmosphere," Donald Johnson, an animal nutrition specialist from Colorado State University, explains the catastrophic effects of long-term global warming and the factors which create greenhouse gas phenomenon. [10]

The average American believes that they are making a difference in the reduction of their carbon footprint by riding a bicycle to work or changing a light bulb, but they still remain oblivious to the fact that switching from a high animal protein diet to a vegan diet is equivalent in carbon reduction to trading in a gas guzzling V-8 for Prius Hybrid. In the production of just one cheeseburger, 10.7 pounds of carbon is created as opposed to just 1.5 pounds of carbon from the creation of a stir-fry vegetable dish consisting of broccoli, carrots, and peppers. Statistically, a family of four can prevent 2,225 pounds of carbon emissions every year simply by just eliminating meat in their diet only once a week!

> *Eating a typical family of four steak dinner is the rough equivalent, energy wise, of driving around in an SUV for three hours while leaving the lights on at home.*

> — (Mark Bittman, Food Matters) [11]

As of 2018, in Brazil, 200 million head of cattle had produced some of the highest concentrations of greenhouse gasses in the world. According to 1988–1995 statistics, compiled by the Brazilian Enterprise for Agricultural Research, the worldwide emissions of methane gas had reached 94 teragrams (Tg = one million tons) annually, with Brazil being responsible for almost 10% of this equation just by itself. On top of all this, burning rain forests to create pastureland further exacerbates this critical scenario. Over 70% percent of all rain forest destruction is for cattle pastures and agriculture, while only less than 3% is for lumber. These fires, which create CO2, represent Brazil's leading source of greenhouse gas emissions. [12]

Methane Emissions Research:

A NASA research study in 2006, clearly documents that the atmosphere was heating up to levels not seen for many thousands of years. In the study led by James Hansen of NASA's Goddard Institute for Space Studies, New York, in conjunction with other major science academia, they overwhelmingly concluded that since 1976, the earth has heated up to levels not experienced in *twelve thousand years!* [13]

By 2018, a team of NASA scientists, led by John Worden, of the Jet Propulsion Laboratory, also documented a disturbing pattern of growth in methane gas production. They attribute this activity primarily to animal agriculture. [14]

The latest studies of satellite photographs in 2021, from Lawrence Livermore National Laboratory, suggest that NASA scientists have been underestimating the extent of global warming gasses for decades! There were previously too many discrepancies in the old data to make it credible. If this new research is, indeed, accurate, then we can expect more severe weather patterns in the future![15]

"The concentration of methane gas in the atmosphere has risen sharply—by 25 teragrams per year — since 2006"

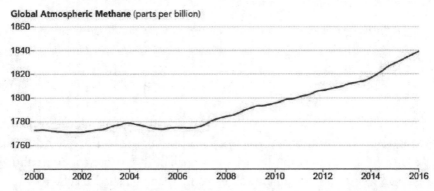

Global Atmospheric Methane (parts per billion)

Under most circumstances, people burn the rain forests and then abandon thousands of herds of cattle on this land in order to show legal ownership. Sometimes cattle are not even the motivating factor in acquiring the land, but rather a necessary part of government politics to qualify for homestead rights. Historically, these types of policies, had deeply concerned many Brazilian environmentalists, such as Rubins Born, head of the nongovernmental group, Vitae Civilis. [16]

Termites: Termites by themselves contribute to methane gas emissions by 400 million tons a year. All this methane breaks down into carbon dioxide in the earth's atmosphere. This figure has become exacerbated by the fact that less than 3% of rain forest destruction is actually for lumber.

Therefore, the remaining rain forest destruction, if not already burnt, is left to rot, subsequently causing mass infestations of termite populations. This worldwide infestation (cattle being the leading cause), is so significant that the total weight of termites on earth outweighs that of man by six to one, while also emitting ten times more carbon dioxide than their human counterparts! [17]

Sustainability:

Eating locally and seasonally grown foods can have a significant impact over the environment. Shipping foods halfway around the world represents a heavy carbon footprint. More than sixteen pounds of CO2 is created by transporting food by air for 5,000 miles in comparison with only one pound or less by ground shipping for approximately 150 miles. The choice is yours; you have real power on your plate! [18]

Notes:

[1] https://www.sierraclub.org/sierra/2019-1-january-february/executive-director/we-need-halt- greenhouse-gas-emissions-while-also?amp

[2] https://climate.nasa.gov/causes.amp

[3] http://www.unenvironment.org/news-and-stories/story/tackling-worlds-most-urgent-problem- meat

[4] https://go.grist.org/e/399522/hane-pledge-climate-index- html/24fypcj/885142983?h=w-K07umkLr4qgwiQqXZAt12AtusSl2QbfFuVD82ybLZg

[5] http://www.worldwatch.org **

[6] Rearing cattle produces more greenhouse gasses than driving cars . . . http://www. un.org/apps/news/story.asp?newsID=20772&CR1=warning.

[7] https://www.facebook.com/OurPlanetbyattn/videos/156154778547294/?t=26

[8] http://enterpriseresilienceblog.typepad.com/enterprise_resilience_man/2009/04/cooking-and- climate.html.

[9] Cow "emissions" more damaging to planet than CO2 from . . . http://www.independent.   co.uk/environment/climate-change/cow-emissions-more-damaging-to-planet-thancosub2sub- from-cars-427843.html.

[10] http://www.tierramerica.net/2000/1126/acent.html.

[11] https://www.scribd.com/book/224282966/Food-Matters-A-Guide-to-Conscious-Eating-with-More-Than-75-Recipe

[12] https://www.theguardian.com/environment/2018/may/10/the-million-dollar-cow-high-end- farming-in-brazil-photo-essay

[13] http://www.giss.nasa.gov/research/news/20060925/.

[14] https://earthobservatory.nasa.gov/images/91564/what-is-behind-rising-levels-of-methane-in- the-atmosphere

[15] https://www.livescience.com/satellites-underestimated-global-warming.html

[16] http://richardsandbrooksplace.org/rubens-born/vitae-civilis-history-and-achievements

[17] http://www.straightdope.com/columns/read/832/do-cow-and-termite-flatulencethreaten- the-earths-atmosphere.

[18] Spotlight: Livestock Impacts on the Environment

The Conscious Planet

We Are the Lightening and the Thunder

We Are the Lightening and the Thunder...
Please Don't Let Them Take our Land and Plunder...
You Know that this is What I Ponder...
And My Heart is Filled with Wonder...
And My Soul is Filled With Fire...
Mother Nature Take Me Higher...
We Are the Lightening and the Thunder...
We Are the Lightening and the Thunder...

Now Remember What I told You...
That the Corporations Bought and Sold You...
You Did What Ever They Told You...
Now There's No time Left to Scold You...
So Just Let Your Inner Light Unfold You...
And Never Let Them Mold You...
Because We Are the Lightening and the Thunder...
We Are the Lightening and the Thunder...

It Would Be Mans Biggest Blunder...
If He Let Nature Go Asunder...
We Would all Be Six Feet Under...
To the Corporations We'd Surrender...
We Are the Lightening and the Thunder...
We are the Lightening and the Thunder...

CHAPTER 9

DUST, DROUGHT, & DESERTIFICATION

The production of Livestock threatens some of the last remaining pristine bastions of unadulterated wilderness left on the planet earth! Raising cattle literally wipes out tropical rainforests. Our rainforests suffer in peril as man's exploitation of the hearty ungulate proliferates. The atmosphere becomes deluged with CO_2, Methane, and other effluvium emissions related to the cattle industry, while the planet heats up and the polar ice caps melt. By the consumption of red meat, humanity has forged a role in the creation of a global environmental crisis!

Delicate ecosystems are contingent upon nature's intricate system of checks and balances. Rainforests are fundamental to all life on earth! This rainforest decimation is further exacerbated by additional land being cleared to grow crops to feed the majority of cattle which are raised in huge commercial, industrial feed lots rather than pastureland.

DESERTIFICATION:

Once taking up over 5 billion acres, 2000 years ago, or 14% of the earth's land surface, tropical rainforests have virtually been cut by two thirds! 57% of this remaining rainforest, is in South America, most of which, in the Amazon region.[1]

A burgeoning demand for land and agriculture, primarily to feed livestock, has placed a severe burden over the earth's ecosystems. Since the year 1800, the expansion of grazing territory has increased by roughly 800%. Deforestation is a contingent factor in the global demand for food in relationship to cattle production. Economic cycles further compound this demand for land and agriculture. Clearing rainforests for grazing land, and to produce grain for feed, has led to irreversible damage to the infrastructure of our precious rainforests. Much of this damage occurs through non-sustainable forest clearing practices, such as the "Slash and Burn" method, most prevalent in the Amazon basin. Between the years 1950 – 1980 more croplands had been created than in the preceding 150 years. In 2006, there were 73.7 million head of cattle in this region as compared to only 26 million head in 1990. [2].

The most significant factor in rainforest destruction from 1960 to the early 90's was not from logging or mining, but from cattle ranching and land speculators who burned huge tracts of rainforest and then turned it into grazing land. Besides Brazil, between 1981-95, Costa Rica,

El Salvador, Nicaragua, and Honduras, also ranked among the top 4 rainforest destroyers from cattle production. [3]

Statistically, 80,000 acres of rainforest are destroyed everyday which is the equivalent to more than one acre per second! In addition, Scientists have also estimated a loss of 137 plant, animal, and insect species per day. (Refer to Endangered Species section)

Even though forests cover about 30 percent of the world's land area, they are still disappearing at an alarming rate! Between 1990 and 2016, the world lost 502,000 square miles (1.3 million square kilometers) of forest—an area larger than South Africa! From the beginning of humanity, 46% of all forest territory had already been destroyed! From a 2015 study in the Journal of Nature, about 17 percent of the Amazonian rainforest had already been wiped out over the past 50 years, and more destruction is currently on the rise![4]

By 2018, these statistics reflected that rainforest deforestation was the "Worst in 10 years!" About 7,900 sq. km (3,050 sq. miles), of the world's largest rainforest was destroyed between August 2017 and July 2018 - an area roughly five times the size of London! [5]

Environment Minister, Edson Duarte, blamed aggressive agriculture production and illegal logging practices. These startling statistics, come amid concerns about the policies of Brazil's newly elected president, Jair Bolsonaro.As a campaign platform, during his bid for the 2018 election, Mr. Bolsonaro had pledged to limit fines for damaging forestry and to weaken the influence of the environmental agency. By late 2019, rainforest destruction was making headline news with fires in Brazil still raging out of control! *Something must be done!* The future of humanity lies in jeopardy![6]

Brazil has the largest commercial cattle herd in the world, (About a quarter billion head), and also being the world's largest producer of soy. (Cattle Feed). Deforestation in Brazil is promoting global environmental change due to the rapid expansion of cattle grazing territories. This rate of growth is so alarming that environmentalists have coined this phrase "Pecuarizacao." (Spanish word meaning "Cattlazation.") This profligate rainforest destruction is further exacerbated by unsustainable pastureland, which is subject to degradation by poor management practices. The final result is cattle that have insufficient grazing territory, further motivating the ranchers to raze (burn), or cut down more rainforest to maintain their herds.

Removing trees inhibits delicate tropical undergrowth from receiving the protection from its rainforest canopy, which blocks the sun's rays during the day and retains heat at night. This disruption has been a contingent factor in temperature swings that can be harmful to plants and animals. Everyone knows that tropical plants need shade. The trees transfer humidity (A system by which plants release water through their leaves.) This moisture evaporates into the atmosphere and thus, creates a cycle of precipitation. For example, 50-80 percent of the moisture in the central and western Amazon remains part of the ecosystem water cycle. Without the proper forest cover, the sun parches the earth, killing the delicate flora underneath and therefore, causing "desertification!" After the 1850's, the conversion of rainforest to pastures and croplands had rapidly expanded in the Amazon basin.[7]

INTERNATIONAL FOOD POLICY RESEARCH INSTITUTE (IFPRI)

Environmentalists at this organization have warned that even a slight increase in land degradation relative to current trends may raise the prices of world commodities from 17 – 30% by 2020. This would automatically result in an increase of child malnutrition along with diminishing food production and food security. This type of scenario would also hinder economic growth, eventually leading to depopulation or geographic disenfranchisement. Going into 2022, we have already seen a significant amount of inflation! [8]

EVERY DAY in this world people eat 21 billion pounds of food

EVERY DAY in this world the animals we raise for food are fed 135 billion pounds ofgrain!!!

EVERY DAY in this world humans use 5.2 billion gallons of water.

EVERY DAY in this world, the animals we raise for food require 42 billion gallons ofwater!!!

EVERY DAY in this world, 80,000 acres of rain forest are destroyed primarily due tolivestock production!!!

If we would raise crops for food rather than for feed then we could help stop Famine, Drought and Desertification of our precious rain-forests!!!

<u>Save the Rainforests!</u>

What would you rather have?
This?   **Or this?**

[9]

Due to rising international demand for beef, with limited land and natural resources, cattle production is already becoming highly industrialized. 80% of the growth in the livestock sector is directly attributed to industrialized factory farming. Cattle are still the *"Single largest Anthropogenic User of Land,"* their production responsible for taking up 45% of all arable land!

This worldwide production of livestock threatens all natural resources on earth; (Air, Water, and Land).[10]

The South American basin has been described by scientists as the Heart and Lungs of the planet, due to its diverse ecosystems and vast water flow from the world's greatest river. 60% of the Amazon basin is contained within Brazil. There are several other South American countries which also occupy this basin.

> *"Calculate the amount of meat eaten by a person in the U.S. per year; translate to numbers of animals. How much energy and grain are used to produce this meat? How many trees in the rain forest are destroyed to produce this meat?"*
>
> — Jeremy Rifkin,

According to FAO reports, grazing territory encompasses 26% of the earth's terrestrial surface, while crops used for feed require 33% of all arable land! These are the main factors in deforestation based on the most recent data available from Livestock's Long Shadow, United Nations Reports. In the Amazon basin, roughly 70% of previously forested jungle is now used for cattle pastures, with a significant part of the remaining land being further degraded by feed crop production. Due to this over production of livestock, some 70% of all grazing land in dryer areas has become severely degraded.

The production of livestock has a major impact over the global water supply. Major factors involve the use of water for irrigation of feed crops. Cattle are also the world's leading source of water pollution stemming from manure, antibiotics, hormones, and chemicals from tanneries, fertilizers and pesticides used in feed crops.

In order to provide huge amounts of cheap affordable beef to a burgeoning U.S. fast food market; major portions of the Amazon rainforest have been sacrificed! You want to feel love and warmth for your family and with these venerations come responsibility and concern for their safety. Now turn this around and look at everything from a macro point of view. Without the trees, the fish in the sea, the birds in the air, and the air and water itself, we could not exist! If people truly love their families and are concerned for their wellbeing, then they would do what is right for them and future generations in making a conscientious decision by recognizing the destructive nature of worldwide livestock production. In 2004, 300 million pounds of beef had been imported to the U.S., just from Central America alone! By 2016, this figure had climbed to approx. 415 million pounds! Many Multi-national food conglomerates including McDonald's and Campbell's Soup, still patronize rainforest beef! [11]

> "The overall impact of livestock activities on the environment is enormous. The health of the environment and the availability of resources affect the welfare of future generations, and overuse of resources and excess environmental pollution by current generations are to their detriment! Ultimately, if left unchecked, environmental degradation may threaten not only economic growth and stability but the very survival of humans on the planet!"
>
> —(Rifkin, Beyond Beef) [12]

FACTORS OF DEFORESTATION:

Clearing Rainforests to Create Grazing Land & Agriculture Production to Feed Cattle

Statistically, cattle ranching, and the production of agriculture to feed cattle, are responsible for 70%*- 91%** (Cowspiracy), of all rainforest destruction; logging only accounts for less than 3%. [13]*[14]**

• Grazing: Cattle over graze this pastureland, they strip the land, eating all vegetation in sight. Their sheer weight from constant perambulation crushes microbes in the soil, prohibiting any new growth. Since major multi-national corporations buy this land so cheap, it is not profitable to re-till the soil and therefore they discard it and buy more land, leaving behind a barren wasteland. Author Jeremy Rifkin, President of The Foundation on Economic Trends, refers to cattle as "Hoofed Locust," describing this phenomenon in his book, *Beyond Beef*, p. 92. This type of "Desertification" is responsible for the majority of all rain forest destruction! Cattle "are also a major source of land and water degeneration, according to FAO report." [15]

• Over Cultivation of Land: The majority of cattle in the Amazon basin are raised in industrial feed lots in huge concentrations. Their demand for grain is high; up to 40 pounds of feed per animal daily. Beyond grazing land, rainforests are also being cleared for the mass production of agriculture to meet the bourgeoning demand of industrial cattle production. E.g. Brazil is the

2nd largest producer of soy, which is primarily used for cattle feed. 98% of soy meal, and 75% of all soy beans produced in the world, are fed to livestock!

• Improper Irrigation Techniques: Improper water conservation practices further reinforce the worldwide drought and deforestation crisis created by the overproduction of livestock.

• Creation of Drought due to the Production of Livestock: Statistically, (taking into consideration all the water used to water crops which are fed to cattle besides the amount of water required in livestock production), it takes up to 2500-8500 gallons of water to produce just one pound of hamburger, as opposed to only 160 gallons of water to create a one pound veggie burger. A complete vegetarian meal with tofu, rice, and vegetables for example, only uses 88 gallons of water.

For example, as was stated in the documentary *Cowspiracy, A Sustainability Secret*, humans use 5.2 billion gallons of water per day, however, the livestock we raise for food consume 43 billion gallons every day! This is conclusive evidence that drought is artificially created by man's greed! Statistically, a dairy farm with only 200 head of cattle can use as much water as a small city! It takes 2000 gallons of water, just to produce only one gallon of milk, while a gallon of oat milk only takes 48 gallons! [16]

Of all water utilized in the U.S., domestic (residential), water use only equates to 5% with animal agriculture responsible for 60% of all fresh water used! [17]

In an article published by author Neil M. Pine: "The Real Truth About California's Drought" on 7/23/2021, at JetSettingMagazine.com, he expounds upon the current corrupt and profligate water conservation policies from California.

In late 2021, during one of the worst droughts in the state's history, Governor, Gavin Newsome, while facing a recall, due to his rampant incompetence involving EDD (CA Employment Development Dept.), fraud, and pandemic restrictions, was afraid to mandate any policies against the real culprits; the Meat and Dairy industry, who OWN him! But rather, in a desperate and abysmal attempt, was trying to shift the blame for this profligate use of water, on to a credulous public, by telling them to cut back their water use by 15%, and never mentioning that 15% would ONLY reduce residential use down to 3.4%, from the already miniscule 4%!!!

> *"California's water resources are being exploited by a handful of powerful ranching interests who are practicing 19th-century water bureaucracy with 21st century sustainability demands!"*

— (Neil M. Pine)[18]

• Will Future Wars Be Fought Over Water Due to Worldwide Drought Conditions?: In a CNBC special which aired 9/23/10 "Water. The Coming Environmental Crisis," it was barely mentioned and never emphasized, how cattle are the #1 cause of drought, famine, pollution, and desertification, and that by becoming a vegan we could help solve this foreboding paradox.

What they also forgot to mention, was that 33% of the entire world's water usage is dedicated toward livestock production. It requires only 10,000 gallons of water per year to produce food for a vegan as opposed to 320,000 gallons a year for average American omnivore. Are vegans eco – hero's or are omnivores merely environmental terrorists?

Fracking uses more than 70 billion gallons of water per year.

Animal agriculture uses 500 times that amount of water.

Anyone omnivore who claims to be a conservationist, ecologist, or say they care about the environment, are nothing more than pure hypocrites!

Due to worldwide drought conditions, climate change and man's encroaching population boom, wars may someday be fought over water. Oil was the predominant factor for war during the 20th century, but water may become the next oil! In his book, *The Epic Struggle for Wealth, Power and Civilization*, Journalist, Steven Soloman, expostulates his belief that water is rapidly becoming a more critical commodity than oil! He claims that only 2.5% of the planet's water supply is fresh. World water usage has increased twice as fast as world population growth! Going vegan is our best solution! [19]

Other major wasters of water are Ethanol and fracking. (Refer to Ethanol chapter)

• Creation of Global Warming: Livestock are responsible for more global warming gasses than all the cars, trucks, and airplanes combined! And in turn, the planet heats up and combined with these other factors, we are literally turning our planet into a desert; a barren wasteland! (See chapter "Cattle and Egregious Greenhouse Gasses.")

We have Sacrificed, Tainted, and Desecrated some of the most Sacred, Pure, and Pristine bastions of unspoiled wilderness left on earth, just so that people can conveniently go to "Taco Hell" or "Shmuck-Donalds!"

> *"An alien ecologist observing earth might conclude that cattle is the dominant animal species in the biosphere."*
>
> —(Jeremy Rifkin)

In the past, many of the U.S. based multi-national corporations had exploited this Latin American Bucolic Bounty: Swift, Armor, Dow Chemical, W.R. Grace, Gulf and Western, Monsanto, and International Foods, were all given assistance from the World Bank, the Inter – American Development Bank, and the United States Agency for International Development. (USAID). As of 2020, ADM, Con-Agra, United Brands, In & Out Burger, McDonald's, Burger King, Carl's Jr, Wendy's, and Cargill Corp, to name a few, are currently active in all this environmental negligence![20]

• Global Cattle Production:

United States: Even more damaging than from mining, recreational vehicles, logging, or energy drilling, grazing still remains as the predominant factor in the creation of pollution, drought, and destruction of biomass in the U.S. In 1934, a major endowment by the U.S. government allowed cattle ranchers access to a majority of public lands for grazing purposes. However, this practice has proven to be ineffective in terms of environmental and economic factors. According to FAO statistics, as of 1992, 70% of all public land in the U.S. was used for grazing! (Approx. 300 million acres), yet with all this sacrifice to our nation's environment, it still only yielded 2% of the nation's beef consumption! As of 2020, there were roughly 94 million head of cattle in the U.S.[21]

"'The Taylor Grazing Act' was accurately described by author Jeremy Rifkin;

> *"Quite simply represented the single biggest give away of land in modern history. No other constituency before or since has been so completely subsidized by the American tax payer, a fact rarely raised in the public debate over welfare programs administered by the federal government."*

> — (Jeremy Rifkin)

• Brazil: In 2006, there were 73.7 million head of cattle in this region as compared to only 26 million head in 1990. Today, there are close to one quarter billion head of cattle, making Brazil, the world's largest livestock producer! In the country of Brazil, as of 2015, only 1% of the population owned almost 50% of all the land! 4.5% of the landowners own 81% of the nation's farmland (Most of it used for cattle production, or growing grain for feed), while 70% of rural households are without land. This country displays more social inequity toward its population than almost any other country on earth! Brazil has the world's worst ghettos with no running water or electricity, referred to as "Shanty Towns" (See chapter "Overfed Cattle = Starving People" [22]

> *"Under Brazilian law, to gain title to public land in the Amazon the land must first be cleared to demonstrate a serious commitment to settlement. Multi – National corporations often wait for peasant farmers to raze and burn forested areas, then purchase the cleared land for a nominal fee to use for commercial pasture. The felled timber is rarely marketed, the colonists finding it more expedient to set it on fire. Astronauts report seeing hundreds of fires twinkling across the Amazon forests in their flybys over the continent. "The commercial loss of timber, though significant, pales in comparison to the ecological cost."*

> –(Jeremy Rifkin)[23]

Finally, in 2006, due to public concern over deforestation, the Brazilian government enacted a law which stipulated that landowner can only clear 20% of their land, therefore ensuring the remaining 80% to maintain forest cover. While this should be acknowledged as a major step toward conservation, there still exists a major ambiguity in the law. Due to discrepancies over land ownership, and lack of clear title, the enforcement of this statute may currently be nonexistent or being incorrectly implemented. As of 2019, 19% of all the land in Brazil was utilized for cattle production. [24]

• Costa Rica: 80% of Tropical rainforest were cleared between1969 – 1989. 2000 powerful ranching families own over half of all the productive land. By 1983, 83% of the beautiful Costa Rican rainforests had been cut down, mostly due to beef production. As of 2019, Costa Rica has 1.3 million cattle. [25]

• Guatemala: Like Brazil, this country also has a ruling class. Less than 3% of the population owns 70% of all the agricultural land, much of it used for cattle ranching. As of 2002, Guatemala had 2/3 of a million head of cattle.[26]

• Honduras: The land devoted to cattle pastures increased from just over 40% to over 60% from 1952 – 1974. During this time beef production tripled! Today Honduras has over 3 million cattle, but there are also many illegal ranches. [27]

• Amazonia: In the region referred to as Matto Grosso, the production of livestock increased from 12 million head in 1994 to 25 million in 2006 and 30.2 million by 2017. In Para, another region of Amazonia, also showed a dramatic increase over the same time period, with production going from 10 million to 20 million head of cattle. By 2017, this figure had increased by more than 50% to 30.2 million! [28]

• Mexico: Famine, deforestation, population displacement, drought, soil erosion, and plague are also direct byproducts of widespread cattlization in Mexico. From, 1960-1980, pastureland in Mexico, increased by 156%. Today, there is an estimated 8.1 million head of cattle.

At that time, 60% of net carbon emissions were attributed to this wasteful conversion of rainforest to pastureland

"We are exporting the future of Mexico for the benefit of a few powerful cattle farmers"

— (Gabriel Quadri, Mexican Ecologist) [29]

> *"The livestock sector emerges as one of the top two or three most significant contributors to the most serious environmental problems, at every scale from global to local ... Expansion of livestock production is a key factor in deforestation, especially in Latin America where the greatest amount of deforestation is occurring."*
> — (Ref: Food & Agriculture Organization of the United Nations. Livestock's Long Shadow (2008)) [30]

Notes:

[1]https://beta.ctvnews.ca/national/climate-and-environment/2021/3/11/15343675.html

[2] (MEA, 2005a)

[3]https://en.m.wikipedia.org/wiki/Deforestation in Central America
https://www.bbc.com/news/world-latin-america-46327634

[4]https://www.nationalgeographic.com/environment/article/deforestation

[5]https://www.bbc.com/news/world-latin-america-46327634.amp

[6]https://mobile.reuters.com/article/amp/idUKKCNISE2IU

[7] Jeremy Rifkin, Beyond Beef; (92:p195)

[8] International Food Policy Research Institute (IFPRI)

[9] Cowspiracy, the Sustainability Secret 2013

[10] Live Stocks Long Shadow Report:

[11|http://www.unep.org/vitalforest/Report/VF6-06-Forests-under-threat-as-agricultural-commodities-take-over.pdf

[12] *Malhi et al. 2008

[13] **Cowspiracy, the Sustainability Secret 2013

[14]http://www.fao.org/ag/magazine/0612sp1.htm

[15] https://www.naturalnews.com/023341_water_milk_organic.html

[16] Cowspiracy, the Sustainability Secret 2013

[17]https://blog.jetsettingmagazine.com/news/the-real-truth-about-drought-in-california-2021

[18]https://mercyforanimals.org/blog/animal-agriculture-wastes-one-third-of-drinkable/

[19]www.npr.org/templates/story/story.php?storyId=122195532

[20]https://earth.org/major-companies-responsible-for-deforestation/

[21]https://www.statista.com/statistics/194297/total-number-of-cattle-and-calves-in-the-us--sim2001/

[22]https://www.npr.org/sections/parallels/2015/08/25/434360144/for-brazils-1-percenters-the-lstays-in-the-family-forever

[23]Jeremy Rifkin, Beyond Beef, (92:p195)

[24]http://www.unep.org/vitalforest/Report/VF6-06-Forests-under-threat-as-agricultural-commodities-take-over.pdf

[25]https://animalsindisasters.org/country-profile/costa-rica

[26]https://www.indexmundi.com/agriculture/?country=gt&commodity=cattle&graph=production

[27]https://www.fws.gov/international/wildlife-without-borders/central-america/honduras-cattle-conundrum.html

[28]https://www.statista.com/statistics/992638/cattle-number-heads-mexico/

[29]https://en.m.wikipedia.org/wiki/Gabriel_Quadri_de_la_Torre

[30] Ref: Food & Agriculture Organization of the United Nations. Livestock's Long Shadow (2008)

The Conscious Planet

Copyright © Neil M Pine 2020

CHAPTER 10

OVERFED CATTLE = STARVING PEOPLE!

*Growing Crops to Feed Too Many Cattle
in a World of Famine*

The production of livestock represents a blatant disregard for the environment (ground, water, and air). It also deprives hundreds of millions of people from third world nations the proper sustenance they need, while at the same time ruining the health of the consumers who patronize it!

There are roughly 1.6 billion cattle on the planet due to man's ongoing lust for dead bovine flesh! Their sheer weight dramatically exceeds that of all human populations combined. All the land they use for grazing combined with all the land which is required to raise crops for feed could be utilized to feed everyone on earth 7 times over! Raising livestock (Either for pastureland or to grow crops for them), encompasses 45% of all the earth's arable land [1]

Have you ever seen a skinny cow in the United States? Of course not! In industrial feedlots, each head of cattle can consume up to 40 pounds of grain or 2 ½% of their body weight daily. In addition, livestock consume 1600% more plant protein than the meat that they yield! (Sixteen pounds of plant protein for every one-pound of meat produced.) [2]

Statistically, humans on this planet eat 21 billion pounds of food per day. However, every day in this world, the animals that we raise for food, are fed 135 billion pounds of grain! The excuse for hunger is a farce when people could be growing grain for food rather than for feed! While thousands of people lay dying of starvation in Ethiopia, millions of pounds of feed crops, produced right there, were being shipped to Europe! Plant-based agriculture provides between five and twenty-six times the amount of protein per any given unit of land than beef, and without the pollution or land degradation. [3]

An increasing level of deforestation remains contingent upon growing populations of cattle and rising demands for grains to meet the needs of industrial feed lots. Over 97% of all soy meal and 75% of all soybeans, produced globally, are fed to livestock while all this grain could feed the starving populations of the world many times over! Due to land restrictions from the overutilization of animal production, Europe must import most of its feed. This grain is primarily derived from developing countries. The irony is that with all this land, cattle, and agriculture, the people of developing countries are literally starving to death! [4]

In South America, indigenous sustenance farmers are forced off their ancestral land and forced to live in squalor, in what are referred to as "shanty towns" (which are the most terrible ghettos in the world, devoid of amenities such as running water or electricity), by corrupt governments who acquiesce with the protocol of multinational corporations like McDonald's, ConAgra and Burger King, who declare imminent domain over the Mestizo Indian land, which once belonged to their ancestors for thousands of years!

These people have no formal education or trade. All they know is how to live off the land the way their forbearance taught them. In most areas of the Amazon region, and many parts of Africa as well, there are no businesses or industries for these people to find an occupation in order to survive. These people literally starve to death, while at the same time, the land that was once theirs is being utilized to grow crops to feed to cattle, and to raise cattle, which will all be sold to the affluent nations of the world! For example, in Brazil, 80% of the land is owned by only 4% of the population, which is primarily used in the production of livestock.

The total meat production in developing countries tripled between 1980 and 2004, while at the same time, record numbers of people from these regions of the world were starving! Mass populations of the world, almost *"1 billion people are starving or severely malnourished.* Out of all these people, (every year) up to 60 million, mostly children, die of starvation or diseases related to malnutrition, while massive amounts of grain are being shipped to the affluent nations of the world! As consumers, there are lifestyle and dietary choices which can affect geopolitical policies. [5] & [6]

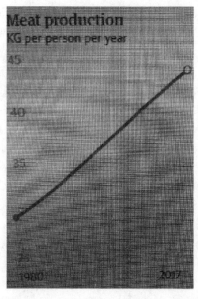

<u>1980</u>   <u>2017</u>

FEED DEMAND FOR LIVESTOCK PRODUCTION:

In April of 2009, the worldwide shortages of wheat, rice, corn, and soy were becoming more prevalent as a bourgeoning third world population pushed commodity prices to record levels! In addition, due to the deadly outbreak of mad cow disease, bovine spongiform encephalopathy, in the 1990s, rendered animal protein is no longer fed to livestock. (Refer to chapter 27 "Mad Cows Englishmen and Howard Lyman.") This has automatically led to increased demand for various forms of vegetable proteins (soy, sorghum, wheat, alfalfa, corn, and other grains or grasses). Due to this and other economic factors, the world soybean production had tripled from 1990–2004. With this rise in grain production, a loss of pastureland will inevitably result. This trend has been projected to increase by 33% in Latin America and the Caribbean, 27% in Sub- Sahara Africa, 6% in South Asia, and 5% in East Asia. (Projection Time Frame: From 1997/99–2030) [7]

This factor will automatically lead to the proliferation of a worldwide industrialized cattle complex, where viral outbreaks and dangerous drug resistant pathogens could more easily become pandemic! [8] See chapter (The Malicious Mechanics of Modern Meat).

In addition, besides the obvious health hazards of factory farming, there also exists two other major negative factors which could substantially impugn any possible benefit resulting from this future expansion of grain production.

The first of which is land degradation, which is a contingent factor in the overproduction of cropland. With this type of scenario, higher rates of ecological damage along with decreased productivity may incur. Increased agricultural production may result in such problems as a buildup of salinity in the soil, water logging, declining soil fertility, increased soil toxicity along with increased pest populations.

The second factor is more complex. In general, it is assumed that the world as a whole displays adequate agriculture production potential. However, by the drought, land scarcity, and poor land suitability, due to cattle ranching, many agricultural models could be failing. Expansion of arable lands into major rain forest biome could result in catastrophic environmental repercussions, which entails the jeopardizing of the earth's ecosystems by creating a loss of biodiversity, water regulation, and erosion control. [9]

Biofuels are also another factor in the loss of sustenance for starving populations. In a starving world, growing food for fuel just doesn't make sense! [10] (Refer to chapter 18, "Ethanol, Government-Subsidized Tax Scam.")

GLOBAL GRAIN PRODUCTION:

Wheat:

China: Increased from only 20,000 tons in 1961 to almost 120,000 tons by 1997, and by 2017, total production had reached 134,000 tons

Europe: Increased from 48,000 tons in 1961 to 85,000 tons in 2001, and over 150,000 tons in 2017.

Latin America: In 2016, South America had produced 29.6 million tons of wheat

India: Increased production from 62 million tons in 2001 to 98.5 million tons in 2017.

United States: The United States has decreased its wheat production from 62 million tons in 1996 to only 47.3 million tons by 2017. [11]

Corn:

China: In 2017, 225 million tons were produced.

United States: In relationship to corn for animal feed, the United States had increased its production by 120% from 1961 to 2001, and by 2017, had achieved a robust production level of 377.5 million tons.

Brazil: 2017: 83 million tons.

India: 2017: 42.3 million tons. [12]

Soy:

Brazil: In 2006, Brazil only produced 52 million tons of soy, at that time, making them the second largest producer of soy beans in the world and accounting for 23% of the world's demand. Approximately 220,000 square kilometers of land in Brazil was planted with soy in 2006. [10] By 2018, Brazil had become the world's leading producer of soy, with almost 119 million tons. [13]

Argentina: This is the world's third leading producer of soy with 40 million tons produced in 2006 and by 2019, the total output for that year is projected to reach 56 million tons. [14]

China: Surprisingly, China is the fourth largest producers of soy, with only 15 million tons as of 2006. [11] In the 2019/20, crop year, China is projected to produce 17.27 million tons. [15]

People from industrialized countries obtain more than 40% of their entire dietary protein intake from animal products, as opposed to 20–30% from less developed countries. [16]

Notes:

[1]   Cowspiracy, the Sustainability Secret http://r.search.yahoo.com/_ylt=A0SO80DNDDB-WucAAAUJXNyoA;_ylu=X3oDMTEzbHByN2Qy   BGNvbG8DZ3ExBHBvcwMxBHZ0aWQDQjExNjBGX-zEEc2VjA3Ny/RV=2/RE=1446018381/RO=   10/RU=http%3a%2f%2fwww.cowspiracy.com%2f/RK=0/RS=kNOzpYhLnD2Ndlf1ET4z0ooBimg-

[2]   www.worldwatch.org

[3]   https://blog.jetsettingmagazine.com/business/war-on-waste

[4]   Latin American Perspectives, Vol. 14. No. 3, Agriculture and Labor: "The Current Food Crisis in Latin America." A Discussion of DeJanvry's "The Agrarian Question," by Timothy Wise (Summer, 1987), pp. 298–315. Published by Sage Publications: http:www.jstor.org/stable/2633683

[5]   http://veg4planet.blogspot.com/.

[6]   http://www.mcspotlight.org/media/reports/beyond.html.

[7]   Pingali and Heisey: Livestock's Long Shadow United Nations Report.

[8]   Refer to Chapter 28 (Swine Flu, Pollution, and Other Health Risks)

[9]   (FAO 2003a): Livestock's Long Shadow United Nations Report. http://www.stockfreeor-ganic.net/index.php?option=com_content+&view=article&id=62:- deforestation-livestock-destroy-ing-the-living-earth-&catid=27:information-about-stockfree- organic&Itemid=56.

[10]   Refer to chapter ? (Subtitle: Ethanol, Government-Subsidized Tax Scam.)

[11]   https://en.wikipedia.org/wiki/International_wheat_production_statistics   http://www.stockfreeorganic.net/index.php?option=com_content+&view=article&id=62:-   deforestation-live-stock-destroying-the-living-earth-&catid=27:information-about-stockfree- organic&Itemid=56.

[12]   https://www.google.com/search?client=safari&channel=iphone_bm&source=hp&ei=Sx-mUXZq5EJ LZ9AP91Y7oBQ&q=corn+production+by+country&oq=corn+priduction+by+&gs_l=mo-bile-gws-wiz-hp.1.1.0i13l8.9806.17922..18941...0.0..0.214.2082.11j6j2......0....1.8..41i13j41j41i131j0j46i-131j0i131j46j46i275j0i22i10i30j0i22i30.937knUng934

[13]   https://www.reuters.com/article/us-brazil-soy-usa/brazil-to-pass-u-s-as-worlds-largest-soy- producer-in-2018-idUSKBN1IC2IW

[14]   https://www.google.com/search?client=safari&channel=iphone_bm&source=hp&ei=mC-CUXe3RB   KS0PEP3467oAg&q=tons+of+soy+produced+in+argentina&oq=tons+of+soy+pro-duced+in+argenti na&gs_l=mobile-gws-wiz- hp.12..33i22i29i30.5117.24961..26333...2.0..0.245.4188.1 7j16j2......0....1.......8. 41j41i13j41i131j0j0i131j46i131j46j46i275j19j0i22i30j0i22i10i30j33i299j33i160.Y_e1URdLdFI

[15]   https://www.reuters.com/article/us-china-crops/china-expects-its-2019-20-soybean-out-put-to- hit-highest-in-14-years-idUSKCN1SG0FR

[16]   (FAO 2006b): Livestock's Long Shadow (Issues and Options). http://frwebgate.access.gpo.gov/cgi-bin/muttidb.cgi.

Be conscious that what you eat can change the ecology, politics, and economic infrastructure of the entire planet!

Become a part of
The Conscious Planet

*We are so skewed
toward efficiency
that we have lost
our sense of humanity.*

— (Jeremy Rifkin)

CHAPTER 11

THE MALICIOUS MECHANICS OF MODERN MEAT

The egregious health and safety standards of the past hundred years, from the days of Upton Sinclair, (author of *The Jungle*), haven't really changed all that much. In fact, in many ways, even with modern USDA standards, the meat today is much more hazardous to the consumer in terms of environmental toxins, chemical additives, and also disease! Also, the employees of these slaughterhouses and factory farms aren't treated much better than they were at the turn of the century! [1]

The slaughterhouse industry preys on economically disenfranchised individuals: immigrants, poor people, and even children. The wages are substandard while at the same time exposing the workers to, statistically speaking, *"the most hazardous conditions of any factory job in America!"* Slaughterhouses also have the greatest employee turnover of any other industry!

Besides the physical danger, the conditions are extremely cruel, loud, filthy, unsanitary, noxious, and malodorous. As is stated in a logo for PETA Organization: *"Meat is Not Green!"* And this is meant both literally and figuratively.

The Department of Labor's Bureau of Labor Statistics reported that one in three slaughterhouse workers suffer from at least one illness or injury annually, as compared to one in ten for other manufacturing occupations. Slaughterhouse employees also have a 3,500% greater propensity of developing repetitive stress injuries than workers of other manufacturing jobs.

Rather than to create safe working conditions by slowing down assembly lines or to invest in better safety gear, these companies would rather cut corners, at the expense of the health and safety of their employees! This industry is notorious for firing employees who have been injured on the job and may be attempting to file a claim.

According to the organization Human Rights Watch ...

"They love you if your healthy and you work like a dog, but if you get hurt, watch out. They will look for a way to get rid of you before you report it. They will find a reason to fire you or put you on a worse job like the cold room, or change your shift so you quit. So a lot of people don't report their injuries. They just work with the pain."

Most factory farms and slaughterhouse operations are set up in the poorest regions of the United States, exploiting masses of underprivileged and uneducated individuals. Slaughterhouse operations often attract illegal migrant workers. These are the type of employees who work very hard for very little pay, and they don't complain for fear of being deported! These big meat producers also exploit children. There are many documented incidents of young teenagers accidentally being killed on the job while working at slaughterhouses. A watchdog group known as the Multinational Monitor calls Tyson Foods one of the "ten worst corporations," due to their egregious record of illegal teenage hiring practices! (See chapter "PETA Organization.") [2]

Covid -19: In late 2021, (WSJ) The *Wall Street Journal*, reported that Covid-19 cases by meat packing workers, were much more than previously known! According to their statistics, the rate of infection among meat packing workers, was 160% higher than what had been reported in 2020! Cases of Covid-19 for 2020, were actually 59,000, as opposed to the originally claimed 22,700. Deaths were also dramatically underreported!

Just another bombshell against the meat industry! [3]

ZOONOTIC DISEASE DANGERS RELATED TO ANIMAL AGRICULTURE:

Adjusted for the cost of inflation, the beef today still costs 30% less than it did in 1970, making it much more attractive to consumers looking for a quick and inexpensive meal. However, even with today's modern safety standards set forth by the USDA, hundreds of millions of pounds of meat have already been recalled since 1988!

Does the average American realize the dangers of the meat they are consuming? Have new federal regulations impugned safety standards for American consumers? As of late 2021, very little has changed from the time of this informative PBS TV special series, *Frontline*, which was aired April 18,

2002, with the topic of the show entitled *"Modern Meat."* They substantiate certain claims about these dangers. Frontline examines the politics involved in the meat processing industry, interviewing current and former U.S. Department of Agriculture officials, meat inspectors, food safety experts, and industry representatives. This one-hour special presentation reveals how the highly industrialized meat industry has compromised safety standards, causing fear and great concern over the spread of dangerous and deadly bacteria in our food chain. Also featured in this documentary are attempts by the powerful U.S.-based Multinational food conglomerates to circumvent federal regulations designed to prevent contaminated meat from ending up in supermarkets or fast-food restaurants. Today, things are no different and even worse! Not only do we still have major meat recalls, but now with Covid-19, this merely compounds the problems! [4]

In our modern Western society today, there exists a lack of accountability on the part of the meat industry.

> *This industry has fought against food safety inspections for a hundred years!*
>
> — (Eric Schlosser) [5]

In the movie *"Modern Meat,"* they examined the entire metamorphosis of modern meat production, from the cattle ranches to the industrial feed lots where up to hundred thousand head of cattle are sequestered until they become fat enough for slaughter. After slaughter, dead carcasses are transported by way of Detroit-style assembly lines. With modern advances in technology streamlining operations, the industrial production of beef carcasses has tripled in some cases, according to industry statistics. Some facilities can process up to four hundred carcasses an hour. However, with this increased efficiency come increased health risks. Animals living in close proximity to one another, sometimes covered with fecal matter, can easily spread viruses and bacteria among themselves. This type of factory farming with pigs was determined to be the cause in the creation of the swine flu epidemic. (See pandemic section of this chapter).

E. coli:

One incident involving E. coli at a Jack-in-the-Box in 1993, motivated the government to seek out a better high-tech system to properly inspect meat. This incident, caused by E. coli bacteria, killed 4 people and injured 750.

Meat inspectors and industry experts have used the standard "poke and sniff" method for decades, visually seeing and physically smelling the carcasses for signs of disease. However, following this Jack-in-the-Box incident, the government had initiated the implementation of a new inspection system referred to as HACCP (hazard analysis and critical control points). This process would require microbial testing to detect the presence of invisible, yet harmful bacteria, such as is found in E. coli and salmonella.

This new form of testing was met with serious opposition from the meat industry. The powerful U.S. food lobbies, which have been heavy contributors to the Capitol Hill lawmakers, aggressively fought against this legislation. In 1996, the USDA started to resist the pressure from the cattle industry and initiated a new inspection system. According to USDA statistics, this new transition to safer standards resulted in a significant decline in the cases of salmonella and other foodborne illnesses. However, don't let this statistic fool you. The

www.NaturalNews.com Created by Mike Adams

U.S. consumer is still in grave danger! The problem with high animal concentrations is the production of feces, which promotes the spread of bacteria and pathogens. The larger the feed lot, the greater the risk of spreading E. coli bacteria and outbreaks of pathogens such as the swine flu epidemic.

Dr. Robert Tauxe, head of the Centers for Disease Control (Foodborne Illness Unit), stated *"The new highly industrialized way we produce meat has opened up new ecological homes for a number of bacteria."* According to Tauxe, you will never get a hamburger patty that came from just one cow. With huge inventories of cattle being herded, fattened, and slaughtered, along with a modern mechanized *"Soylent Green"* style of assembly-line production, it would be impossible to determine just how many cows were involved in the production of just one burger! Tauxe goes on to say that *"if we take meat from a thousand different animals and grind that together, we're pooling bacteria from a 1,000 different animals as well."*

Also due to the unsanitary living conditions on factory farms, dangerous new forms of bacteria are created.

> *Escherichia colia O157:H7 is a relatively new strain of the common intestinal bacteria (no one had seen it before 1980) that thrives in feedlot cattle, 40 percent of which carry it in their gut. Ingesting as few as ten of these microbes can cause a fatal infection; they produce a toxin that destroys human kidneys.*
>
> — (Michael Pollan) [6]

Salmonella:

Salmonella is the most common cause of food poisoning. Food poisoning affecting about 1 in 6 people. It originates from fecal matter, contaminating meat and vegetables. This contamination of vegetable crops is all contingent upon animal agriculture!

Due to unsanitary living conditions, more than 65,000 soldiers died from salmonella during the civil war. In 1985, Contaminated dairy products from Hillfarm Dairy, infected more than 16 thousand people and killed at least 10!

In 2015, cucumbers imported from Mexico infected 907 people in 40 states. There were 200 hospitalizations and 6 deaths! [7]

Every year, thousands of cases of food poisoning are reported according to the CDC (Center for Disease Control). And at the same time, it maintains a running list of tainted meat recalls. While international trade has dramatically increased, so has the risk of diseased cattle coming from foreign countries, threatening U.S. livestock populations. For example, USDA inspector, Roger Viadero uncovered 650,000 pounds of foreign meat which came from a country that was embargoed because of foot-and mouth disease. This meat got into the United States somehow, and who knows how much comes in that they don't find?

Another move by the corrupt cattle industry to circumvent safety regulations, was to create legislation to give the courts authority to limit government enforcement of these new safety standards. Case in point, in the TV special "*Modern Meat*," they mention a lawsuit filed by a Texas meat grinding company, Supreme Beef, against the USDA. The USDA had effectively shut the company down after it failed to pass bacterial contamination tests on three separate occasions! In one incident, test results confirmed that almost 50% of Supreme Beef's meat was found to be contaminated by salmonella.

The company still sued the USDA, stating that the government didn't have the right to shut down its operations simply due to its failing to meet USDA standards. Also supporting Supreme Beef in this lawsuit was the National Meat Association. The outcome of these legal proceedings was nothing less than scandalous! The Federal Appellate Court ruled in favor of the meat industry. This author sincerely hopes that for this appellate judge and his family that, like Oprah, *"they never eat another burger."* There once was a time when civil servants cared about the citizens they served. The meat industry has helped to reduce our society to corruption, wickedness, and disease!

According to Carol Tucker Foreman, head of food safety at the Consumer Federation of America and former USDA official, modern meat production is putting consumers in danger and leaving them vulnerable to widespread outbreaks of bacterial contamination. She refers to a case of deadly Listeria connected to the Ball Park Franks brand, where sixteen deaths and five still births were reported. [8]

> *"These hot dogs were shipped everywhere, and thousands and thousands of them were made every day. So the potential for one mistake rippling out and causing thousands of deaths is there."*

> — (Carol Tucker Foreman)

Pandemics:[1]*

Historically, due to filthy, overcrowded conditions on industrial feed lots, such as factory hog farms, or in exotic meat markets*, virulent strains of infectious diseases can proliferate. When will humans finally learn? All these global pandemics stem from raising animals in cruel and unhealthy conditions. (Going vegan would eliminate these problems) [9]

§ **Avian Flu: (H7N9)l**

In February of 2013, the Avian (H7N9), became an active contagion, transmitted from chickens to humans. Around 100 people were infected, killing almost 20% of them. This virus was traced to several types of fowl (Mainly chickens and ducks). [10]

---

1   This chapter originated form an article published in JetSettingMagazine.com in March of 2020. At that time, it was assumed that like every other pandemic throughout history, this one also came from animal agriculture. However, since late 2019, NO animal has ever been linked to this virus! More than 80,000 wild and domestic animals have already been tested! And what a COINCIDENCE, the virus broke out in Wuhan China, in the same city that has a viral laboratory where weapons grade coronavirus testing was being conducted! (Refer to chapter "Shots of Ignorance")

§   **SARS: (SARS-Cov)**

A strain of the Corona Virus referred to as SARS, was also related to Asian, exotic animal, meat trade. The Civet, A small cat like creature, was directly linked to this pandemic, lasting 6 months and causing worldwide havoc with almost 9000 cases and 774 deaths. [11]

§   **Corona Virus: (Covid19)**

As of 9/20/21, the fatalities of this virus had surpassed the Spanish Flu, making Covid-19, the deadliest pandemic in history with more than 675 thousand deaths worldwide! [12] *

§   **Swine Flu: (H1N1)**

As we know, too many animals, living in close proximity created the H1N1 Swine Flu virus. 9/11/09: In an article conducted by investigative reporter, Scott Harris; it is explained how this pandemic was directly linked to factory hog farming in Mexico.

At that time, health experts had concluded that the origin of this disease was linked to Carrol Farms, a huge industrial hog farm located in the city of Perote, in the state of Veracruz, Mexico. Carrol Farms is a subsidiary of U.S. based, multi-national, Smithfield Food Corporation. In March of 2009, Mexican health officials issued an alert, claiming that 60% of the residents of Perote were suffering from symptoms associated with this outbreak. [13]

POLLUTION, AND OTHER HEALTH PROBLEMS RELATED TO FACTORY FARMS:

In the last fifty years, the number of ranches used for animal production has decreased sharply. However, during this same time span, the number of animals used for food production has risen. Therefore, this has inevitably led to greater and greater concentrations of animals living in close proximity to one another, and it is this overcrowding which is believed to have caused the Swine Flu epidemic in 2009. This type of IFAP, (Industrial Farm Animal Production), has been linked to many other out breaks over the years. [14]

A philanthropic watch-dog group known as the PEW Commission in Washington D.C., has been monitoring such conditions since 2005. On 9/29/08, they released their report on IFAP. In this study, the egregious health risks of factory animal production are discussed. *"Commissioners have determined that the negative effects of the IFAP system are too great and the scientific evidence is too strong to ignore."* Animals kept in close confinement along with industrial feeding methods have led to increased pathogen risks which have dramatically increased the propensity of transferring the disease from animals to humans. [15]

Furthermore, people who reside near these industrial fed lots are subjected to noxious air emissions, which have the greatest effect on the young, elderly, and people with acute pulmonary or chronic heart disorders.

Another serious health issue due to negligent IFAP practices, is the over imbuement of animal waste into the soil. *"As with public health impacts, much of the IFAP's environmental impact stems from the tremendous quantities of animal waste that are concentrated on IFAP premises."* This abnormal concentration of waste inevitably works its way into the environment, poisoning wells, underground streams, rivers, and lakes. It is also the runoff of this waste which has been directly linked to E-coli contamination. (Refer to Chapter 5, "Pollution") By no longer patronizing these Pernicious Pork Purveyors (in the eating and slaughtering of animals), we not only avoid pandemics and contaminations such as this, but we also intrinsically become more humane individuals, while at the same time, we dramatically reduce our carbon footprint and prevent chronic illness.

MAJOR RECALLS:

The Centers for Disease Control (CDC), in conjunction with the Federal Department of Agriculture (FDA), regulate and promulgate all food recalls. Large scale recalls can be devastating to a company's bottom line, affecting millions of people from all over the country, potentially costing millions of dollars in losses from lawsuits, health care costs, product destruction, sanitation upgrades and loss of consumer confidence.

- Cargill Corp:
  In 2011, one of the greatest poultry recalls, was for 35 million pounds of ground turkey This salmonella contamination was responsible for 1 death and 75 people sickened. After all this, the plant was only shut down for 1 week before reopening. [16]

- **Hallmark/Westland Meat Recall:**
  One of the largest beef recalls in history took place in California in 2006. An incredible 145 million pounds of beef had to be destroyed. Animals were horribly mistreated and diseased! An undercover video showed cows too sick to walk, being loaded up to go to the slaughterhouse. A third of this meat was for school lunches! [17]

- **JBS Meat Packing:**
  2018, the world's largest meatpacker recalled 6.9 million pounds of beef linked to antibiotic- resistant Salmonella. The meat originated from the JBS meat packing plant in Arizona, causing 120 illnesses and encompassing 22 states! [18]

Dangerous Meat Additives:

- **Nitrites: Nitrites** are used in the preservation of meats, primarily for preserving botulism by inhibiting the growth of bacteria which causes it. Food editor for the *Today Show*, Phil Lempert stated that *"nitrates combined with amines naturally present in meat to form carcinogenic N-nitroso compounds."* When cooked, these carcinogenic compounds increase the risk of stomach, pancreatic, and colorectal cancers. [19]

In another research study published by Center for Public Health Practice Organization, they discovered an alarming rate in the incidents of brain cancer in children of women who ate large amounts of cured meats while pregnant. [20]

- **Recumbent Bovine Growth Hormone (rBGH):** The hormones injected into livestock also have deleterious effects on both cattle and human populations. These hormones simply make cattle very fat, so what do you think it does to people? In 1994, the FDA secretly approved of the genetic growth hormone referred to as recumbent bovine growth hormone (rBGH). At that time, the Monsanto Corporation had already invested 500 million dollars in the creation of this hormone to increase milk production. Inevitably, this growth hormone has created suffering and disease for the cattle and to the humans who ingest it. During the time of this introduction to rBGH, dairy farmers were beginning to report that their cattle were becoming sick with a disease called mastitis.

To fight this disease required massive dosages of antibiotics. Due to this reason, 95% of dairy farmers initially refused Monsanto's pernicious protocol. However, due to the powerful influence that Monsanto had on the agricultural industry, all these ranchers finally gave in to industry pressures. This bovine growth hormone also increases the production in humans of another powerful naturally occurring growth hormone referred to as IGF-I, which has also been linked to cancer! During this period, a scientific voice of reason was needed. Besides such prolific crusaders in this field like Jeremy Rifkin and Howard Lyman, Robert Cohen was also instrumental in the fight against Monsanto and rBGH. He publicly remonstrated his belief not only about the collusion between Monsanto and the FDA but also as to a major cover-up by the scientific community! His three years of research proved him to be alarmingly correct! In addition, his research also uncovered a connection between Monsanto, the FDA, and congressional leaders! Cohen revealed conclusive scientific evidence that laboratory animals treated with rBGH developed cancer. However, even with all this evidence, Cohen could still not convince the FDA to ban the use of this hormone. As usual, America has been sold out to the big corporations! [21]

• **Diethylstilbestrol (DES):** This chemical is primarily used for meat processing. Eighty-five percent of all meat in the United States contains harmful residues of this chemical poison additive! Besides displaying carcinogenic properties, this chemical has also been linked to artificially speeding up the process of puberty in girls. Many countries from around the world refuse to import meat from the United States for this reason. [22]

• **Antibiotics: Super Bugs: (Antibiotic Resistant Strains of Infections Disease)**

As if there weren't already enough reasons to go vegan, then this one is the clincher!

Another major abuse of the food chain, is the meat industry's widespread use of antibiotics. It was recently reported that the incidence of Super Bugs (Antibiotic Resistant Strains of Infections Disease), could reach epidemic proportions, possibly killing up to 50 million people around the world, by 2050! Cattle are routinely loaded up with these drugs to counteract the spread of infectious bacteria due the close confinement and filthy living conditions of factory farming. The startling statistic is that 80% of ALL antibiotics manufactured in the United States is fed to livestock! This practice is directly linked to strains of human bacterial infections which are dangerously resistant to antibiotic treatments. [23]

• **Transglutaminase (Moo Glue?):** Restaurants and food producers use this enzyme ingredient to form steaks out of stewed chunks of meat. This ingredient is also used to make imitation crab and chicken nuggets. While the main ingredient in meat glue might not be hazardous to your health, what it enables meat producers to do with it is another story. All these discarded scraps of meat could contain harmful levels of bacteria which may contaminate each other. This is why it is always recommended that a person who ingests substandard beef products always cook their food very well done. [24]

• **Ammonium Hydroxide** ("Pink Slime"): McDonald's recently made an official public announcement that they have discontinued the use of "pink slime," a dangerous chemical additive in their hamburger meat. This public disclosure by McDonald's coincidentally came on the heels of statements made by popular TV chef Jamie Oliver, who blew the whistle on McDonald's and other corporations using this controversial chemical. Of course, coincidentally ShmuckDonald's claimed that their decision to discontinue the use of "pink slime" (ammonium hydroxide), was "not related to any event." YEAH RIGHT! Like one day it just dawned on them *"Hey, our burgers are already slimy enough without having to add extra slime to them."* Who knows how many million Bags of Slime that McDonald's has already surreptitiously mixed into their hamburger meat over the years?

Many countries from around the world have banned the use of this hazardous chemical substance. But despite McDonald's recent actions, the FDA still allows U.S. corporations to use this "scurrilous dreck." Chef Jamie Oliver stated that rejected or discarded beef trimmings, which are used for dog food, were being treated with "pink slime" to kill high levels of bacteria and then were being recycled back into hamburger patties.

This chemical is found in household cleaners, fertilizers, and is similar to the chemical used as an explosive by former terrorist Timothy McVeigh, to blow up the federal building in Oklahoma City. Maybe McDonalds's was also concerned that their own customers might have explosive flatulence, or do they all just smell that way from eating there? Furthermore, have you ever wondered what burned holes in your shorts every time you ate at Mickey D's? So, while ShmuckDonalds may no longer use "Bags of Slime," then this should not mitigate the fact that they are still "Slime Bags!" [25] & [26]

Notes:

[1]   www.capitalcentury.com/1906.html.

[2]   www.peta.org/issues/animals-used-for-food/more-reasons-to-go-vegan.aspx.

[3]   https://www.wsj.com/articles/covid-19-cases-in-meat-plants-were-much-higher-than-previously- known-report-says-11635371970

[4]   movies.msn.com/movies/movie-synopsis/frontline-modern-meat.

[5]   www.downtr.co/895936-eric-schlosser-fast-food-nation.html.

[6]   Michael Pollan, The Omnivore's Dilemma: A Natural History of Four Meals.

[7]   https://www.ranker.com/list/biggest-salmonella-outbreaks-in-us-history/eric-vega

[8]   https://consumerfed.org/expert/carol-tucker-foreman/ [9]

        https://www.googleadservices.com/pagead/aclk?sa=L&ai=DChcSEwiit5K93dDzAhWp-
GK0GHYUVBD0YABAAGgJwdg   &ae=2&ohost=www.google.com&cid=CAESQOD2_V6VjbDslTg-
ZIOGeZGgmCZ5rlEydc4Zd3rND5LLtMiwOC4OCpch5vBXeco_BzH5228v8yAgrlh_O0Gsszz1Q&sig=A-
OD64_0hEh-   O3Ua2FjuHSSjQeZ1mhJWBoQ&q&adurl&ved=2ahUKEwiHr4q93dDzAhXOCTQIHdy6B-
3MQ0Qx6BAgEEAE

[10]   https://www.googleadservices.com/pagead/aclk?sa=L&ai=DChcSEwig2oCX-   PPzAhXh-
H60GHbwoCkEYABAAGgJwdg&ae=2&ohost=www.google.com&cid=CAESQOD2R_NZ-Wfy4NCx-
jltC-   3TjdpchDT-kigd99WX6diRlyTOxWiElxwz4sSQpthfRN35T8v1q_1Z37aRBGELjYv4&sig=A-
OD64_1y7ar4XqCy06-   ftV6k7eEGUm59yw&q&adurl&ved=2ahUKEwisrPiW-PPzAhW7FTQIHSF-
pB38Q0Qx6BAgGEAE

[11]   http://www.cidrap.umn.edu/news-perspective/2004/01/who-sees-more-evidence-civet-
role-sars

[12]   https://www.usnews.com/news/national-news/articles/2021-09-20/us-coronavi-
rus-death-toll- surpasses-1918-flu- pandemic#:~:text=The%20U.S.%20coronavirus%20death%20toll,-
from%20the%201918%20flu% 20pandemic.&text=For%20a%20country%20of%20roughly,have%20
died%20from%20the%20co ronavirus.

[13]   https://www.cdc.gov/flu/pandemic-resources/2009-h1n1-pandemic.html

[14]   http://www.ncifap.org/_images/PCIFAPFin.pdf

[15]   https://conversationinfaith.com/tag/pew-commission-on-industrial-farm-animal-produc-
tion/

[16]   http://www.cnn.com/2011/HEALTH/08/03/turkey.recall/index.html

[17]   https://www.yourlawyer.com/food-poisoning/recalls-food-poisoning/usda-hallmarkwest-
land- beef-recall/

[18]   https://www.foodprocessing-technology.com/news/jbs-tolleson-recalls-6-9-million-
pounds-beef- products/

[19]   www.ehow.com/facts_7194274_danger-nitrate-meat-preservatives.html.

[20]   https://r.search.yahoo.com/_ylt=Awr9DtrQPX5hQYgA5OdXNyoA;_ylu=Y29sbwNncT-
EEcG9zAzY   EdnRpZAMEc2VjA3Ny/RV=2/RE=1635692113/RO=10/RU=https%3a%2f%2fwww.publi-
chealthp ractice.org%2fnutrition-and-cancer/RK=2/RS=ryn92XR6Rp3EWnglgtveZlktf.k-

[21]   https://www.cancer.org/cancer/cancer-causes/recombinant-bovine-growth- hormone.html

[22]   https://www.jstor.org/stable/2683843

[23]   Super Bugs https://www.bbc.com/news/health-30416844

[24]   en.petitchef.com/recipes/meat-glue--fid-1422045.

[25]   news.yahoo.com/blogs/sideshow/mcdonald-confirms-no-longer-using-pink-   sli-
mechemicals-171209662.html.news.yahoo.com/blogs/sideshow/mcdonald-confirms-no- longer-us-
ing-pink-slimechemicals-171209662.html.

[26]   slimeblog.chron.com/ . . . /2012/01/mcdonalds . . . pink-slime-in-burgers.

"It's a Slime of the Times"

So go Vegan

and become a part of…

The Conscious Planet

*"The Doorstep
to the Temple
of Wisdom
is a Knowledge
of our own Ignorance"*

– C. H. Spurgeon, *Gleaning Among the Sheaves*

CHAPTER 12

MAD COWS, ENGLISHMEN AND HOWARD LYMAN

Howard Lyman has been a crusader for environmental health awareness for more than 2 decades. Commonly known as "The Mad Cowboy," he is the cattle rancher who went vegan, gaining his notoriety on the Oprah Winfrey show, where after disseminating information about the dangers of beef production, Oprah was quoted to have said, *"I'll never eat another burger!"* This statement triggered a 12-million-dollar lawsuit filed against Oprah and Mr. Lyman by the Texas Cattlemen's Association. However, based on the preponderance of evidence, the court ruled in favor of Oprah and Howard Lyman, because in the court's own words "You can't sue someone for telling the truth!" (Refer to chapter "The Psychology of the Cattle Culture") [1]

According to Howard Lyman, there exists a strong relationship between Alzheimer's disease and Mad Cow disease, (Bovine Spongiform Encephalopathy) (BSE). In 1907, Dr. Alzheimer published a medical treatise outlining the documentation about a disease that he would someday be named after. During his research, he had initiated the help of two assistants; Dr. Creutzfeldt and Dr. Jacob. These two doctors later identified a similar brain wasting disease which was accordingly referred to as Creutzfeldt Jacob's disease, but today is more commonly referred to as Mad Cow disease. What they discovered was a disease which attacks the central nervous system, thus causing massive brain degeneration, and usually initiating from a long incubation period. The outcome of which is always fatal! This disease is transmitted through the consumption of infected body parts, especially the brains and spinal tissue.

Once the meat is infected with this disease, no amount of cooking can destroy this dangerous element. Even after exposure to over 1000 degrees

Fahrenheit, antibiotics, bleach, boiling water, along with a slew of solvents, detergents, and enzymes, (which are traditionally utilized to destroy most commonly known bacteria and viruses), the infectious agent in the Mad Cow disease still remains alive! I guess they weren't kidding when somebody said "Don't Mess with Mother Nature!" [2]

> *"Mad Cow disease is important today, not just as a deadly food borne illness, but also as a powerful symbol of all that is wrong about industrialization of farm animals."*
>
> — *Eric Schlosser [3]*

> *"Meat is a lush medium for Pathogenic Bacteria and Germs; it can harbor Parasites, Toxic Chemicals, and Metal Contaminants. And now it can bring death by brain rot!"*
>
> — Steve Bjerklie [4]

During the late 90's the UK began to be plagued by incidents of (BSE). This started a new resurgence of awareness about Mad Cow disease, where new laws were passed internationally, safeguarding against the spread and transmission of this disease. Howard Lyman was instrumental toward the international eradication of this disease, and also helped to educate ranchers and consumers as to the risks and dangers associated with (BSE). [5]

It was revealed by Mr. Lyman that the USDA knew as early as the late 1980's, of the risks and dangers associated with this disease, but did nothing to stop it! Documentation from early USDA reports substantiates that a strict ban on "Animal Cannibalism" (*"The feeding of rendered slaughterhouse waste from cattle to cattle as protein and fat supplements"*), would have to be initiated in order to prevent the spread of (BSE). But, due to the corrupt nature of our capitalistic system, the cattle industry, (with all of their powerful lobbying and special interests) refused to support such a ban due to economic feasibility. [6]

Today it is highly acknowledged throughout the livestock industry and the world that you should never feed meat or the byproducts of such to herbivorous animals. Well Duh; you think that might have been a hint when these animals were classified by science as Herbivores in the first place? Some countries even issue the death penalty to any rancher caught feeding animal ruminants to cattle. Man had to learn the hard way what Mother Nature had already known for 4 billion years!

Notes:

[1]   http://www.prwatch.org/prwissues/1997Q2/lyman.html

[2]   http://www.notmilk.com/m.html

[3]   Eric Schlosser, Fast Food Nation, The Dark Side of the All American Meal: (01:272)

[4]   Steve Bjerklie, Executive Editor, Meat Processing Magazine 2/27/06

[5]   Rachel's Environment and Health Weekly, July 9 and July 16, 1998 [02.06.27:01]

[6]   www.madcowboy.com/01_FactsMC.DefHis.html

Remember: Cows would never get Mad in the first place if we didn't eat them!

Make one happy today; go Vegan and become a part of...

The Conscious Planet

Copyright© 2020 Neil M. Pine

CHAPTER 13

SHOTS OF IGNORANCE
Boost Your Immune System Naturally!

COVID-19 VACCINES CAN BE DANGEROUS AND INEFFECTIVE!

In an article published by author **Neil M. Pine**, in March, 2020 (Global Pandemics All Linked to Animal Agriculture in JetSettingMagazine.com), had "assumed" that all the factors involved in the creation of the Covid19, Swine Flu, SARS. and the Avian Flu had been identified. [1] Historically speaking, without man's negligence and greed for dead animal flesh, these pandemics could never have proliferated.

However, new evidence is mounting that the original outbreak of COVID-19, may be MAN-MADE and came from a laboratory that Dr. Anthony Fauci, funded through his organization, the National Institute of Health (NIH), In Wuhan China! Since late 2019, no animal has ever been identified as being the source of the outbreak, with over 80 thousand animals tested so far (both wild and domestic). [2]

Big Pharma, in collusion with the World Health Organization (WHO), the Center for Disease Control (CDC), and the Food and Drug Administration (FDA), along with the mainstream media (MSM), is once again capitalizing on a seemingly credulous public by instilling a false sense of paranoia! They're the ones who make it so dangerous, poisoning people with their filthy vaccine, building panic into the equation, telling people what they want them to know, not what they should know! One of the biggest lies told by the drug companies, through the (MSM), is that the vaccine will actually protect you from catching the virus! Yes, the virus is dangerous, however, what they're not telling you, is that there are completely safe, non-toxic, dietary regimens, which represent the quintessential method towards boosting the immune system! This powerful regimen may be able to prevent Covid-19 or dramatically mitigate its symptoms, . (We will discuss this further at the end of the chapter)

Due to the mutating variant of the virus, the efficacy of the vaccines has become impugned! As the virus mutates, the antibodies may become ineffective in stopping the spread of pathogens. The shot developed for a particular strain of Covid-19 may become ineffective by the time a person receives it.

Contingent upon this shortcoming, people are being warned to take a 3rd and even 4th shot! This is all pure insanity! Even more alarming, are the recent studies that indicate that these vaccines may be "cytotoxic," presenting an elevated risk of danger!

Further evidence suggests that the vaccine doesn't just stay in the area of the inoculation, as it was intended, but rather, it spreads throughout your entire body; to the brain, heart, lungs, and digestive system, thus triggering reports of palsy, heart attacks, fever, strokes, blindness, and even death! And since the CDC "claims" that the vaccine is safe, then all these dangerous side effects are being blamed on other conditions and therefore, never get reported by the media!

The false assumption is that the spike protein is innocuous, but thorough testing was never completed!

According to Dr. Robert Malone (Inventor of RNA vaccine technology), "the spike protein is very dangerous!; it's cytotoxic!"[3] The Salk Institute, founded by polio vaccine pioneer, Jonas Salk, has declared a RED FLAG WARNING on the Covid-19 vaccine! The current 4 vaccines available are designed to inject the spike protein directly, or via RNA technology. According to the Salk Institute "The spike protein introduced to the vaccine has resulted in seriously high rates of adverse reactions, including strokes, blood clots, heart attacks, palsy, blindness, and even death!"

In their latest report in mid-2021, they state that "The SARS-CoV-2 spike protein, present in the vaccine and virus, is creating serious vascular damage in addition to respiratory illness!" The spike protein "damages cells" and causes "vascular disease" even without the virus! This research study clearly demonstrates an in-depth understanding of the dynamics involved with the spike protein and its effect on the vascular system.

"The vaccines were designed to contain the very elements that are killing people" "All Covid vaccines should be immediately halted and recalled"
— (Salk Institute) This is all caused by the spike protein that's deliberately engineered into the vaccines!

Thousands of people have already succumbed to this gross malfeasance against humanity! [4] [5] In 2019, Dr. Sherri Tenpenny was interviewed on the Coast to Coast AM Radio Show with host George Noory. Dr. Tenpenny is a naturopath and clinical nutritionist who is regarded as an international expert on vaccine injuries and problems associated with vaccines. She has been working in this field for more than 17 years. She is a board-certified Emergency Room (ER) physician and was formerly a director of an ER for 12 years. In 1996, she started an integrated medicine practice. Since that time, people from all 50 states of the U.S., and 17 foreign countries, have all sought her help.

In her seminars, she explains the insidious pseudo-science of vaccines along with the wanton collusion between big pharma and the medical industry! She goes on to say that these so-called virulent strains of disease (such as chickenpox and measles) are actually just common infections, not viruses, that all children should get in order to build up their immune systems later in life. Therefore, vaccines for these diseases are just another scam by big pharma! To mitigate or prevent these infections, she recommends vitamin D supplements.

She goes on to explain that the dangers of vaccines have been well documented by all the mainstream medical literature, such as the Center for Disease Control (CDC), *The Journal on Pediatric Infectious Diseases*, and *The New England Journal of Medicine*.

Pharmacists and doctors are well aware of these statistics, but since the drug companies subsidize their salaries, then this automatically creates a conflict of interest! Out of 100 people who take a vaccine, about half experience negative reactions, some getting very sick, paralyzed, and even dying! As Dr. Tenpenny likes to say "Figures don't lie, but liars do figure!" Even the Supreme Court ruled that "Vaccines are unavoidably unsafe"

The vaccine business is a total fraud! In modern medicine, double-blind studies are considered the gold standard. However, in the vaccine industry, these types of trials are never implemented. Instead, these criminals attempt to obfuscate the truth with their own bogus clinical trials which falsely claim the vaccine to be as safe as a placebo.

However, in reality, the placebos that they use are actually another vaccine or even a very toxic dose of aluminum. In this way, they can say that the results of the vaccine were not much different than the placebo, which is a total lie, because the true meaning of this word placebo should always represent an innocuous, inert or neutral substance, not another dangerous toxic substance. [6]

It is the strong belief of many scientists, including experts in the field of microbiology and virology, that these dangerous vaccines were actually developed in collusion, with the same CRIMINALS who created the virus itself!!!!!!!!!!!!!!!!!!!!!![7]

How ingenious! And while the world is overcrowded anyway, then we can knock off a few % of the population and make billions of dollars for the drug companies, all at the same time! Don't believe it? It was documented that Moderna patented the Covid-19 vaccine in Sept. of 2019, COINCIDENTALLY, right before the virus was "released!"[8]

It was man's ignorance that created this disease and it is even more ignorant to use this deadly vaccine to deal with it! There are vaccine commercials every 10 seconds on the radio, brainwashing the public as to the safety concerns and doing everything possible to appeal to all aspects of society! Young, old, gay, straight, white, black, etc… Laws have been passed making it mandatory for many workers in the government and private sectors to get the vaccine!

Obviously, if the drug companies sponsor airtime and advertising space, it's hard to tell the public any negative news about them for fear of losing advertising dollars. Even in a depressed economy, drug companies are one of the only industries which seem to be impervious to recession. The media desperately needs their revenue stream. Pharmaceutical companies are among the most powerful corporate entities on earth.

So, therefore, how can the media deny their avaricious protocol? This qualifies as a classic case of government and corporate subreption (A malicious concealment of the truth), for that mighty dollar – which in this case is definitely – the root of all evil!

Today, the pharmaceutical industry is out of control. In the year 2000, the annual worldwide vaccine revenue was only 5 billion dollars. The Covid-19 vaccine is estimated to reach 157 billion In sales by 2025! [9]

In 1986, the Vaccine Injury Compensation Program went into effect. Since that time, over 3.2 billion dollars in compensation has already been paid out to thousands of victims. The purveyors of these vaccines claim that they are safe. So then, why have they already paid out so much money?

Over the years, many vaccines have been banned! For example, the vaccines Gardasil and Cervlorex were banned in 7 countries around the world. They have been linked to chronic long-term disabilities and even death! With all the documented toxic side effects, how many more vaccines can a person take? So, therefore, it should be considered very disturbing that there are currently 54 vaccines already on the market with another 310 in development. [10]

In a 6 countrywide study, it was concluded that a primarily plant-based diet, not only significantly reduced the risk of catching COVID-19 but also dramatically mitigated the dangerous symptoms associated with the disease!

Participants in this research project displayed a 73% reduction in the odds of contracting moderate to severe COVID-19, compared with the group on a standard omnivore diet! [11]

Even after all this solid evidence, some people are so brainwashed by the media, that they actually still think the vaccine is good for them. These are the same type of people who stare into microwave ovens all day and look like the Pillsbury Dough Boy!

And it's no wonder.... An article published by Healthgrades.com. "10 Things You Need To Know about Coronavirus," is supposed to be an informative, state of the art expose', supposedly giving people critical advice in dealing with this pandemic. Yet, not one of these recommendations mentions the most critical factor for the prevention and treatment of any flu: Boosting your immune system through nutritional therapy. This article is so typical of this ignorance by the MSM! [12]

IMMUNE SYSTEM BUILDING DIET W/HERBS & SUPPLEMENTS:

THIS AUTHOR TAKES 4 FLU SHOTS A DAY!

4 one-ounce shots of Wheatgrass juice!

A vegan macrobiotic, with a raw juice diet, including vitamin and mineral supplements, is the quintessential method to protect yourself from the virus and boost your immune system to its MAXIMUM capacity!

Would you play with fire around gasoline? Then why would you drink milk during a flu epidemic? Drinking milk or using other dairy products during a viral pandemic makes you a "Covid Kamikaze!" If you wish to dramatically increase your chances of catching this current variant, then go out and consume a lot of ice cream and milk every day!

Viruses thrive on the congestion and bacteria created by dairy products. What kills 90% of Covid-19 victims? Pneumonia! Where the lungs fill with fluid! So why not go up to a Covid patient who can barely breathe, and then offer them a big cold glass of Phlegm, I mean milk! (Refer to chapter "Dairy; Dirty, Dangerous, Dysfunctional, and Disgusting")

So, the best advice to start off with is to refrain from dairy products, especially during a viral outbreak. In doing so, a person not only does themselves a great favor, but they also show respect for others by not spreading their filthy germs everywhere as a result of their dietary ignorance. [13]

This author boasts about taking FOUR FLU SHOTS A DAY! What is meant by this hypocritical sounding predilection, is that he drinks 4, one-ounce shots of wheatgrass juice every day, along with other raw juicing practices Wheatgrass juice contains anti-neoplastic properties which have been proven to prevent and treat many forms of cancer. These powerful healing properties come from the live enzymes, amino acids, and the live oxygenation factor to your bloodstream, which also supercharges your immune system, reduces the risk of allergens, and thus, can prevent the contraction of contagions, with no harmful side effects.

Wheatgrass is considered by nutritional experts to be an excellent source of vitamins A, B, C, D, & E, and K, which also plays a critical role in our immunity.

So let's go through this one more time:

Wheatgrass fights and prevents cancer; the vaccine causes it!
Wheatgrass boosts your immune system; the vaccine degrades it!

Wheatgrass eliminates or lessens the severity of allergens, the vaccines cause them! Vaccines are highly toxic; Wheatgrass is proven to eliminate toxins from our cell tissue!

Yet, if you ask your doctor about Wheatgrass, then they would probably say "what's that?" [14] "In addition, immune system boosting herbs and supplements, such as Ginger, Moringa tea, Vitamin C, Oregano oil, NAC (N-Acetyl-L-Cysteine), Colloidal Silver, Horseradish, Zinc and zinc throat Lozenges,

Pure Quinine, Chaga mushroom extract, Vitamin D3 (Vegan), Golden Seal and Echinacea, work extremely well in the prevention and treatment of cold and flu symptoms. Remember, it's important to drink plenty of fluids, (dairy-free), 70% of which should be water." [15]

And to take this prevention even one step further, one should try to exercise a vegan macrobiotic diet. No animal products (meat or dairy), fried foods, processed grains or sweeteners. (Organic and free from chemical additives would be the superlative choice). [16]

Notes:

[1]   https://blog.jetsettingmagazine.com/environment/global-pandemics-all-linked-to-animal- agriculture

[2]   https://www.technologyreview.com/2021/03/26/1021263/bat-covid-coronavirus-cause-origin- wuhan/

[3]   https://youtu.be/Du2wm5nhTXY

[4]   http://www.sensation-hub.com/2021/05/bombshell-salk-institute-science-paper.html?m=1

[5]   https://www.salk.edu/news-release/the-novel-coronavirus-spike-protein-plays-additional-key- role-in-illness/

[6]   https://naturopath4you.com/vaccine-101-with-dr-sherri-tenpenny/

[7]   https://brandnewtube.com/watch/dr-david-martin-dr-reiner-fuellmich-july-9-   2021 RlmKScwsMf6ATEG.html

[8]   https://blog.jetsettingmagazine.com/drugs/dr-anthony-fauci-and-the-greatest-crime-in-modern- history

[9]   https://www.reuters.com/business/healthcare-pharmaceuticals/world-spend-157-billion-covid- 19-vaccines-through-2025-report-2021-04-29/

[10]   https://naturopath4you.com/vaccine-101-with-dr-sherri-tenpenny/

[11]   https://nutrition.bmj.com/content/early/2021/05/18/bmjnph-2021- 000272?fbclid=IwAR-3LaC9e6rXm0pdmXmOaDC_dNVFwUUR0JIr1ul8QGS-xh9n9j-QnKKAZm1E

[12]   https://www.healthgrades.com/right-care/coronavirus/10-things-to-know-about-coronavirus

[13]   Go Dairy Free: Is your Immune System Confused?http://www.godairyfree.org/200712042150/News/Nutrition-Headlines/Is-your-Immune-System-Confused.html

[14]   Health Banquet: Wheatgrass Juice Studies Prove Effectiveness of This Marvel http://www.healthbanquet.com/wheatgrass-juice-research.html...

[15]   https://blog.jetsettingmagazine.com/news/what-is-the-state-of-the-art-lifestyle-part-2

[16]   http://www.godairyfree.org/200712042150/News/Nutrition-Headlines/Is-your-Immune-System- Confused.html

So Don't Be a Sucker for Big Pharma!

Become a part of …

The Conscious Planet

CHAPTER 14

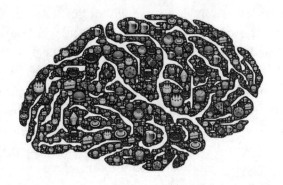

THE RELATIONSHIP OF DRUG ADDICTION TO FOOD CRAVING

"By 3 Methods May We learn Wisdom
First by Reflection,Which is the Noblest
Second by Imitation, Which is the Easiest
& Third by Experience Which is the Bitterest!"
—(Confucius)

Special Note: In regards to the brutal and devastating drug fentanyl, users from around the world are overdosing at alarming rates! Fentanyl overdose deaths in Los Angeles County are up more than 1200% since 2016, according to the L.A. County Department of Public Health. From 109 in 2016 to 1526 as of 2021! The U.S. Drug Enforcement Administration states that this deadly synthetic opioid is 50 to 100 times stronger than morphine! The information from this article could absolutely save someone's life[1]

The use of junk foods or fast foods creates biochemical imbalances in the brain, potentially causing addictive personalities. This seems especially true for young and developing minds. Many children are raised on sugary and saturated fat-based products. If you watch a children's television show, you will notice that half the commercials are advertising sugary snacks, low in fiber and wholesome nutrition. Besides obesity and childhood diabetes, these high-sugar and fat diets lead to behavioral disorders such as Attention Deficit Disorder(ADD), and hyperactivity, which may result in anxiety, and/or depression later in life.[2]

The type of personality created suffers from a greater propensity towards addiction to drugs or alcohol. Alcohol turns to sugar in the bloodstream and drug addicts crave sugar. Eating junk food is a "quick fix" mentality. If someone uses drugs or alcohol habitually, they are considered by society to be endangering their health and psychological well-being. What society doesn't seem to notice is the daily bombardment of chemicals: caffeine, nicotine, artificial flavors, artificial colors, preservatives, food stabilizers, anti-caking agents, pesticides, hormones, and high levels of saturated fats and sugar.[3]

There is no wonder why so many people go through phases of binging and purging, along with anxiety or depression. These behavioral swings can result in violent or even suicidal tendencies in many individuals.[4]

Physician and author, Dr. James P. Gills, states that "Criminals, both juvenile and adult, consume high dosages of soda and dairy products." Also, spousal abusers tend to eat large quantities of meat.[5]

Bernie Goetz "The Subway Vigilante," who was acquitted of shooting four youths in self-defense, ran for mayor of New York in 2002. His campaign platform was based on vegetarianism, stating that a diet free from animal products could help to quell inner city violence as well as police brutality. If elected, he vowed to mandate a vegetarian menu in all government buildings, including all schools, libraries, hospitals, jails, and other government facilities.[6]

Over the years, most schools around the nation have banned soda and junk food from their vending machines, replacing them with juice and more healthy snacks.[7]

The Muse school in Callibasses, Ca, is the first primary and secondary school in the nation to go Vegan. The school has also implemented a "Seed to Table" program, which offers "a hands-on way for students to learn about the connection between their everyday food choices, their health, and the health of our environment," [8]

Michelle Perro, one of the top pediatricians in the nation, has warned parents about GMO ingredients in their children's food! She explains that glyphosates in the form of Roundup herbicide, have been linked to autism, ADD and brain fog in children! It not only affects the mitochondria but also damages the microbiome, being associated with the leaky gut syndrome and fatty liver disease! In the past, children were never diagnosed with fatty liver! This is a red flag for parents.[9]

Health and fitness enthusiast, Jack LaLane, wrote an autobiography that explains that as a teenager, he tried to murder his own brother on three separate occasions. He attributed his violent tendencies to a high-sugar and fat-based diet and claimed he changed his whole personality by no longer indulging in these foods.[10]

There is also the precedent court case known as the "Harvey Milk Decision," where it is claimed that due to the overconsumption of Twinkies, Dan White became temporarily insane, and was driven to murder two elected officials. The jury agreed; thus he was acquitted. This is infamously referred to as the "Twinkie Defense," but as we all know, it was due to a drug called sugar![11]

Other side effects associated with the use of sugar include heart disease, obesity, tooth decay, diabetes, and hyperglycemia.

Take, for example, the use of hard drugs; the principal reason that people become addicted to these substances is due to the damage incurred to their nervous system. Even short-term use may permanently impair the nervous system. This damage automatically creates the craving, and long-term abuse not only creates intense cravings but can also seriously affect the internal organs.

Our body considers any form of inebriation from drugs or alcohol as toxic. The kidney and liver are overworked as the organs try to purge the system of these poisonous chemical substances, in an attempt to not induce damage. But prolonged drug use forces the body to transform and adjust, thus we develop a tolerance to these foreign chemical substances, while at the same time we are damaging our internal organs.

The nerve damage associated with the abuse of drugs and alcohol is created by the inability to function of bio-chemicals that naturally transmit information from one neuron to another. These neurotransmitters become damaged by factitious chemicals (drugs and alcohol), so the nerve synapses will no longer properly produce these natural biochemical substances. As these neuro-receptors become clogged and damaged, the production of natural hormones, such as endorphins, which reduce the sensation of pain, becomes blocked.

E.G.: In an attempt to become sober after a prolonged period of drug abuse, there might be an excess of certain types of enzymes present in the body, and a deficiency of other types of neurotransmitters. The body is not ready to re-establish its original balance and therefore goes through stages of withdrawal.

Drugs such as Prozac, Xanax, Methadone, etc., merely subdue the craving for controlled substances, yet do nothing to heal the nervous system or internal organs, and may have serious side effects, such as allergic reactions, psychological problems, and even further damage to the kidney and liver. In addition, the long-term effects of psychotropic drugs may include permanent psychosis, making a person into a zombie, worthless to society, a ward of the state, an S.S.I. recipient, another statistic that the taxpayers will be responsible for.

No wonder members of NA and AA proclaim that they are still drug addicts or alcoholics after many years of sobriety. This is due to the fact that they still suffer from permanent damage to their nervous systems. The craving is still there, just waiting to be triggered, and tobacco, coffee, and donuts, (nicotine, caffeine, and sugar), which are prevalent at these meetings, are not going to heal them. If anything, the use of these alternative substances would make it more difficult to deal with the problem of chemical dependency.

To further substantiate this assertion, here are several quotes from Pharmacological M.D., Darryl S. Inaba, in his book entitled *Uppers, Downers, All Arounders.*

"Since withdrawal from prolonged amphetamine use is accompanied by physical and mental depression, the majority of patients who want to stop can be treated by encouraging abstinence, not drug substitution, and by intensive counseling. Users should avoid all stimulants, Including caffeine and tobacco."

> *"Nicotine: Tobacco is the most addicting drug there is. Nicotine craving, in fact, may last a lifetime after withdrawal."*

> *"The most dangerous psychoactive substance in the United States is nicotine. Anywhere from 300,000 to 400,000 of the country's 56 million smokers die each year from heart disease, emphysema, or cancer, all consequences of smoking. This drug is legal and over the counter."*[12]

Lobbyists of the omnipotent tobacco industry have successfully fought to keep tobacco legal, despite the fact that federal law prohibits the sale of any products proven to be carcinogenic. [13]

Caffeine: Coffee is considered a mild stimulant when taken in moderation; it produces an effect of alertness and helps give us the impression of being more awake and less fatigued. The side effects include: An increased heart rate, raised blood pressure, an irritated stomach, nervousness, disorientation, irritability, muscle spasms, and insomnia. With long-term abuse, it is possible to become afflicted by coronary heart disease, ischemic heart disease, heart attacks, non-malignant lumps in women's breasts, diabetes, liver problems, cancer of the pancreas, and calcium deficiency associated with osteoporosis[14]

KREB CYCLE:

Caffeine does not give the body nutrients to produce energy. in the form of ATP: (Adenosinetriphosphate), It merely blocks the Adenosine receptors that are signaling that the body needs rest! The Kreb cycle utilizes nutrients from carbohydrates, fats, and proteins that release energy into our bodies.

Caffeine provides an artificial form of energy without providing the nutrients that create real energy! This may also disrupt your natural sleep cycle![15]

Why not call coffee what it really is: "Caffeine and glucose laden, pancreas festering, diabetes-inducing, osteoporosis juice!" * (Assuming that a sweetener was added) If this author were ever asked, "What's one of the best ways to prematurely age?" Drinking lots and lots of coffee every day would be a good answer. So when you're strolling up to your "Hoity-Toidy," upscale, coffee house, then just keep on walking another 100 feet or so, there should be a juice bar somewhere around there. If not, this shows what a bunch of unconscious cretins the local business people in your area really are. However, if you're an entrepreneur with a vision of the future, then this could be your golden opportunity. Open your own juice bar! The people drinking all that coffee are killing themselves. The people who go to juice bars are in it for the LOOOONG haul! They won't be kicking the bucket any time soon like those overweight, jittery, ratchety, hyper-tense, prematurely-aged, caffeine junkies!

The reason for this facetious attitude is primarily due to the ludicrous nature of our society. Here we have an upscale coffee house with a bunch of lazy, but hyped-up slobs, exercising their jaw muscles, when they could all really be doing themselves a favor, with a caffeine-free, truly invigorating and revitalizing, wholesome and pure, Raw Juice Experience!

Modern-day food and medicine are defiling the health of the American public, thus destroying immune systems, internal organs, and psychological well-being. People feel sick, so they turn to drugs or alcohol. The proliferation of toxins through modern-day living is culminating in our body, affecting our mind in the process. Drugs and alcohol merely fill the void created by the ups and downs swings of unhealthy living.[16]

By exercising the system of macrobiotics outlined in this chapter, we can heal our nervous system, rejuvenate our internal organs, and find peace of mind. Because, by purifying the body, we become more sensitive to our external environment, and in turn, we are affected more easily by toxins from insalubrious food products, as well as drugs. Once your system is purified, the wrong substances, (junk food, drugs, or alcohol), will make you ill.

Naïve opinion would have you believe that you're making yourself oversensitive. What they don't realize is that the body is healing itself, rejuvenating itself, back to the original state that nature intended, and by sensitizing your system, you also become desensitized to your cravings. When you purify your body, you may experience a greater sense of spirituality, a natural high that nothing can replace. Now you can focus better, sleep better, and feel more refreshed. You no longer crave the substances which led to your addiction, whether it's from junk food or drugs. You truly are free!

The contingent factor of craving-type foods, like drugs, is that they are all partial or modified from the original state that nature created. The word unwholesome means not whole, an altered representation of its natural form, and like drugs, these foods have a diverse molecular chemical structure that stimulates your mind in various ways. When we remove the hull from brown rice or bleach whole wheat flour, we are removing the natural whole goodness of the food. The essential distinction between whole and processed foods is the overall effects on the brain. Unprocessed foods contain a greater spectrum of natural chemicals than processed foods. These varieties of substances in natural foods counterbalance themselves, so they do not radically affect the brain. Insalubrious food products have an aberrant effect over their natural counterparts. In removing the basic wholesomeness of food, the chemical effects on the brain become intensified by taking away this natural balance. Processed substances like sugar and white flour have very distinct chemical structures.

The more incomplete a food is, the more likely it will resemble a drug, rather than a measure of sustenance. Heroin, cocaine, and sugar are all processed from plant material. We only develop cravings for processed foods, and due to the chemical effects on the brain, we use these substances like drugs rather than for their food value.[17]

Like drugs, emotional craving for food falls under two areas of classification: 1: Anxiety. 2: Depression. When we are depressed we may feel down, lethargic, or fatigued; we are looking for a quick "pick me up." With anxiety, we feel nervous, irritable, or wound up, we are looking for some way to "mellow out," relax, or just have some peace of mind. Foods that we crave only temporarily relieve the condition of this basic dichotomy. What ends up happening is that after we consume these substances to deal with our condition, we inevitably end up in a worse frame of mind after we "come down" from the effect. Our metabolism is "put through the ringer," and like drugs, we crave more of the substance to get us back on track; that fifth cup of coffee, another soda, a whole box of cookies or donuts. Unwholesome foods that we crave never deliver the peace of mind that we're looking for, and can greatly inhibit our physical, spiritual, and psychological well-being.[18]

The propensity toward spiritual acumen through macrobiotics supersedes all factitious stimuli. By practicing this system, we can achieve calm restfulness, complacency, and peace of mind, avoiding depression. We can simultaneously feel more alert with synchronized brain waves, better reaction time, memory enhancement, increased concentration, and spiritual awareness, without anxiety.

Just as we cannot be close to God by using drugs or alcohol; too much of the wrong foods can also inhibit our spiritual growth. "Our Body Is Our Temple," yet we treat it like a toxic waste dump! Truly mystical experiences can be gained by the purification of the body through macrobiotics.

In the Bible, the moral and ethical implications of an insalubrious diet are discussed in Daniel. However, some theologians would still remonstrate that according to Biblical terminology, God blesses all food put on earth for man. But when the Bible was written, our food chain did not contain processed, genetically and biologically altered, irradiated, chemically preserved, artificially flavored and colored, pesticide imbued, and antibiotic-tainted products.

"But Daniel made up his mind that he would not defile himself with the king's choice of food or with the wine which he drank; so he sought permission from the commander of the officials that he might not defile himself."

— *(Daniel 1:8)*

In the story of King Nebuchadnezzar, Daniel chose a simple diet of vegetables instead of meat and other rich foods of the king. At the end of the story, King Nebuchadnezzar was sentenced by God to eat grass in a field like an animal for seven years. According to the Bible, this was God's way of purging his spirit of evil. This story is metaphorically synonymous with the way that wheatgrass juice purges your body of toxins, When you read between the lines, the message becomes strikingly clear: A PURE DIET EQUALS A PURE SPIRIT![19]

Many self-help programs like NA and AA rely on spirituality to cure addictions. They offer excellent psychological support and counseling and are all very noble causes, but they still lack the critical components of a macrobiotic diet. However, in the Buddhist 12-step program, it "takes the treatment to a whole new level," by practicing the principles of Ahimsa (feeling compassion for all living creatures), it naturally becomes easier to refrain from eating meat. Therefore, these ascetic qualities, based on humanitarian principles, can ameliorate our spirit. This intrinsic strength of will and spirit can help serve as a critical factor within the framework of our sobriety. Unlike traditional 12-step programs, they treat the addiction as spiritually inhibiting, rather than referring to it as a physical disease. The focus is on the mind, rather than on the body, to free us from our addictions.[20]

In the book, *Professor Arnold Ehret's Mucusless Diet Healing System*, published 90 years ago, he extols the virtues of raw foods free from animal products. According to Professor Ehret, all mental illnesses could be cured through a vegetarian diet without processed sweeteners.[21]

All bad habits and negative behavioral conduct becomes easy to understand when you realize that they are merely efforts to temporarily feel better. People must get in touch with their inner selves to overcome their addictions. This inherent spirituality is becoming more difficult to obtain as our culture and way of life tend to discount the spiritual experience; hence, in this society of fast women, fast cars, and fast food, people are looking for the "quick fix," and are turning to drugs, alcohol, and junk food, to fill the hollow feeling in their lives.

Here is a quotation by Mathew Fox, from his book entitled The Coming of the Cosmic Christ, quoted in the *New York Times* book review, Jan. 15th, 1989:

> "A civilization that denies the mystic is no civilization at all. It offers no hope and no adventure, no challenge worthy of sacrifice and joy to its youth or artists. It offers no festivity, no Sabbath, and no living ritual to its people. And no deep healing. Such a culture promotes negative addictions: Drugs, crime, alcohol, consumerism, and militarism. It encourages us to seek out outside stimulants to provide meaning for life and defense from our enemies because it is woefully out of touch with the power inside".
>
> — (Mathew Fox)[22]

Just by itself, a vegan macrobiotic diet is an important factor in the control of our cravings, but it is this complete holistic system, which can significantly help us to overcome our addictions.

Genita Petralli, a licensed, board-certified H.H.P., N.C., and M.H. is the author of *Alcoholism: The Cause & The Cure*. This powerful book utilizes nutritional therapy to help people overcome their addictions![23]

Linus Pauling, Nobel Prizewinning professor and founder of the Linus Pauling Institute, has demonstrated prolific success in regard to this paradigm:

> "Orthomolecular medicine is one that aims to restore the optimum ecological environment for the body cells by correcting the imbalances or deficiencies on the molecular level based on individual biochemistry using natural substances such as vitamins, minerals, amino acids, enzymes, hormones, and essential fatty acids."
>
> – (Linus Pauling)[24]

Nutritional therapy can help us to strengthen the nervous system while at the same time, rejuvenating our internal organs.

For Cocaine and Methamphetamines:

- Magnesium: (Mineral), This mineral will curb the craving for alcohol and narcotics, as well as, nicotine, caffeine, and sugar.

- Vegan sources: Wheatgrass, pumpkin, sunflower, flax seeds, walnuts and almonds, tomatoes, various beans and legumes, spirulina, and other dark green vegetables, (refer to diet section).

- L-glutamine: (Amino Acid), Reduces sugar cravings which are prevalent with substance abuse.

- Tyrosine: (Amino Acid), A building block for the depleted neurotransmitter dopamine.

- D-phenylalanine: (Amino Acid), A compound of adrenalin and the neurotransmitters noradrenalin and encephalin.

- Nor epinephrine: (Neurotransmitter), People who use cocaine become deficient in this substance.

- Live, raw juices.

- Vegan macrobiotic diet

For Heroin and Opiates:

- Magnesium supplements: (refer to 1A).

- Clonidine: (Amino Acid), Stress will deplete endorphins in the nerve cells, causing a person to rely on heroin to stimulate the action of the missing neurotransmitters in the emotional centers of the brain; the area which gives the user a sense of well-being. Clonidine dampens the withdrawal symptoms of opiates, and also offers effective detoxification.

- L-glutamine: (Amino Acid), (refer to 1A)

- Live, raw juices (refer to diet section).

- Vegan macrobiotic diet

For Tobacco:

- Magnesium supplements, (refer to 1A)

- Live, raw juices.

- A vegan diet: By becoming vegan a person's blood will turn more neutral to alkaline, instead of acidic. This in itself will reduce the craving for nicotine by up to 40%.

For Alcohol:

- Magnesium supplements (refer to 1A).

- Kudzu: (Herb), This herb boasts miraculous healing effects. Chinese physicians have known for centuries that Kudzu actually helps to heal internal organs damaged by alcohol while suppressing the craving.
- Vegan macrobiotic diet

For Sugar:

- Cinnamon
- Ginger
- Wheatgrass juice
- Vegan macrobiotic diet

ADAPTOGENIC HERBS:

Consists of a group of herbs that may prevent hormonal imbalances resulting from stress, thus creating a state of homeostasis while helping people to better cope with stress by the following:

- Protecting the liver
- Improving blood/sugar metabolism
- Inhibiting craving for alcohol and controlled substances
- Enhancing stamina and energy
- Building muscle tone
- Enhanced healing properties
- Providing more energy for focus and concentration
- Decreasing anxiety
- Promoting sleep and relaxation
- Helping people become more productive
- Instilling a sense of well-being

List of Adaptogenic Herbs:

Ginseng

Ashwagandha

Astragalus

Licorice Root

Schisandra

Jiaogulan

Reishi mushroom

General information:

Natural supplements for discomfort associated with substance abuse, and as an alternative to psychotropic drugs. Use accordingly in conjunction with therapies from sections 1-4:

• SAM-e: S-Adenosyl Methionine: (Amino Acid): Has a contingent factor in at least 35 biochemical processes in the body. Recent studies have documented some very effective antidepressant properties. SAM-e increases levels of serotonin, dopamine, and phosphatides.

• Melatonin:(Hormone) Can be successfully utilized in the control of sleep patterns associated with withdrawal. It also contains elements that are known to reduce the severity of depression in cases involving people who live in climates that don't adequately receive enough sunlight.

• L-Tryptophan:(Amino Acid), It is a precursor in the production of serotonin which calms us down and helps to induce sleep. The effects of L-Tryptophan are quite synonymous with the consumption of turkey. Everyone knows eating turkey makes you sleepy; however, vegan supplements are recommended.

• Kava Kava:(Herb), For relaxation and the promotion of sleep.

• Valerian root: (Herb), Like Kava Kava, Valerian root also displays soporific effects.

• Skull Cap:(Herb), For pain.

• Lachesis:(Herb), For suppressing emotion.

• Ignatia:(Herb), For the treatment of shock.

• Gelsemium:(Herb), Conditions involving grief or shock.

• Causticum:(Herb), For depression, (feeling negative or gloomy).

• Metallicum:(Herb), For despair or guilt due to loss or failure.

• St. Johns Wart: (Herb), Excellent for the treatment of depression associated with withdrawal.

• Marijuana:(Herb), It is a plant that affords many legitimate medicinal properties; an excellent non-toxic source for pain relief, insomnia, depression, appetite stimulation, and bipolar disorder. *Cannabis may have negative effects on some substance abuse patients depending on the type of substance abuse and stage of addiction. However, they would be so much better off if they could stick to marijuana!

• Psilocybin mushroom: Microdosing this psychedelic mushroom has been proven effective in cases involving PTSD, Bipolar disorder, and substance abuse! The mushroom itself is completely non-addictive and also affords health benefits, such as high fiber while providing vitamin D, and niacin. thiamine, riboflavin, biotin, cobalamins, and ascorbic acid. [25]

Without the proper nutritional therapy, a person may have sobriety, but not a full recovery

Disclaimer: Everybody's body chemistry is different. Certain medical conditions may preclude some remedies so always try to seek professional advice before committing to any system. These are only general recommendations. What works well for some may have detrimental effects on others. Besides consulting with a holistic health care professional, each individual should experiment with various regimens, (Dosages & combinations of herbs, vitamins, etc.), which may achieve optimum health results.

Notes:

[1]   https://www.instagram.com/reel/Cll1WrsDuFC/?igshid=YmMyMTA2M2Y=

[2]   Encyclopedia of Junk Food and Fast Food by Andrew F. Smith Publisher: Greenwood (August 30, 2006) ISBN – 10: 0313335273

[3]   Alcoholism: The Cause and the Cure, the Proven Orthomolecular Treatment, by Ginita Petralli: Licensed, board-certified Holistic Health Care Provider, Nutritional Consultant, and Master Herbalist. https://www.amazon.com/Alcoholism-Cause-Cure-Genita-Petralli/dp/1591965101

[4]   Dr. Abravanel's Anti-Craving Weight Loss Program: 1/1/91, by Elliot D. Abravanel

[5]   www.lovepress.com/gills.htm

[6]   http://www.berniegoetz.net

[7]   OCA Joins Nader Org. to Ban Junk Food in Schools: http://www.organicconsumers.org/Organic/schoolfood.com.   [8]   https://vegnews.com/2019/2/how-one-all-vegan-california-school-plans-on-changing-education-forever

[9]   https://drmichelleperro.com/

[10]   www.imdb.com/name/nm0482364

[11]   en.wikipedia.org/wiki/Twinkie _defense

[12]   Uppers, Downers, All Arounders: Physical and Mental Effects of Psychoactive Drugs, by Darryl S. Inaba, Pharm. M.D., and William Cohen, 6th Edition ISBN: 978-0-26544-28-4

[13]   http://www.cdc.gov/tobacco/campaign/tips/resources/data/cigarette-smoking-in-unit-

ed-states.html

[14]	Uppers, Downers, All Arounders: Physical and Mental Effects of Psychoactive Drugs, by Darryl S. Inaba, Pharm. M.D., and William Cohen, 6th Edition ISBN: 978-0-26544-28-4

[15]	https://youtu.be/rr7IRYLqleg

[16]	https://macrobiotics.org.uk/vegan-is-trending/

[17]	The Four Whites-A True Story by Craig Lane
http://macrobiotics.co.uk/articles/fourwhites.htm

[18]	Ibid

[19]	The Bible: [Daniel 1:8], the Story of King Nebuchadnezzar

[20	http://the12stepbuddhist.com/cure-for-addiction-attachment-gone-wild

[21]	https://www.amazon.com/Arnold-Ehrets-Mucusless-Healing-System/dp/0990656403

[22]	https://www.amazon.com/Coming-Cosmic-Christ-Healing-Renaissance/dp/0060629150

[23]	https://www.thedrpatshow.com/guest/genita-petralli,1222.html

[24]	https://www.alternativetomeds.com/services/holistic/orthomolecular-medicine/

[25]	Claudia DeSantis (MS) (CNS) Www.PlantFullPleasures.com

Mr. Pine's work is endorsed by …

Nutritionist: Claudia Desantis, MS, CNS
Professor of Child Development: Dr. Carol Sigala PhD
Holistic Physician: Dr. Zayd Ratansi ND

CHAPTER 15

DAIRY
Dirty, Dangerous, Dysfunctional, and Disgusting!

The idea of drinking cow's milk has been ingrained in the psyche of the American public for hundreds of years. The big lie is that this insidious, chemically imbued, mucus-causing substance, rife with disease-causing bacteria and harmful saturated fats and cholesterol, actually has some health benefits? Everybody needs milk, right? Just ask gold medalists, Mark "Spits" and Peggy "Phlegming," superstar crooner Frank "Snotra," or one of the most prolific physicians of all time, "Mucus" Welby, MD!

The truth is that nobody needs cow's milk except for a baby calf! The only people who still drink this slimy disgusting cow snot are brainwashed by their own doctors and/or the National Dairy Council. (Milk is accepted and endorsed by the American Medical Association (AMA) under the current dietary guidelines in the National Food Pyramid). [1] No wonder *"dead doctors don't lie."* If all these poor naïve people could only read English or any other translatable language, then they would become privy to scientific documentation which verifies the egregious health effects of this scurrilous dreck! Even Dr. Spock, who was probably one of the most prolific pediatricians of all time, clearly stated that *"cow's milk was only intended for baby cows."* And this statement was made more than sixty years ago, before the additional modern dangers of hormones, antibiotics, radiation, and pesticide residues were known. What is so significant is that he was a pioneer of veganism! [2]

According to statistics published by USDA, milk and dairy products make up a significant portion of the American diet. In 1995, the average American consumed 584 pounds of milk, cheese, butter, sour cream, ice-cream, etc.; 394 pounds of vegetables; 193 pounds of meat; 192 pounds of flour; and only 121 pounds of fresh fruit. As is stated in the book *Milk the Deadly Poison,* *"That's one very lopsided food pyramid!"* However, when considering all the gross and disgusting forms of adulterated dairy products represented, then this author would more appropriately refer to this food pyramid as "Slopsided"! [3]

The National Dairy Council disseminates "lie after lie," year after year, until finally through time, the public becomes brainwashed to their avaricious propaganda! On their home page they show a doctor in a white coat. [4] It is their primary purpose to promote dairy products at all costs. They have to help get rid of the millions and millions of gallons of this hideously toxic, wretched, cootie-infested glop, before it all spoils and becomes even more contaminated and toxic than it already was to begin with! It is this author's humble job to educate the public so that hopefully "it does all spoil!" Thankfully, due to this literature and many other publications like it, the public is becoming more aware, and consequently, the entire dairy industry now faces bankruptcy! [5]

Statistics indicate that by the time the average American reaches the age of fifty-two, they have already ingested enough fat from milk to be equivalent to consuming one million strips of bacon! Obviously, fat and cholesterol are factors in today's obesity crisis in America, but did you know that the increased use of antibiotics in dairy cattle are also major contributors to this epidemic?

So, the idea of staying fit and trim by drinking milk from a *Cow* is just more *Bull* from the Dairy Council!

Lie # 1, Bone Health: One of the first major lies is that milk is good for the bones. However, research conducted as early as 1955, substantiates evidence that when you pasteurize milk, you also destroy key enzymes which help assimilate calcium into your bones. Cow's milk contains inflammatory properties! It was never designed for human consumption, containing about 3 times the protein as human milk, which creates metabolic disturbances in the human body that have detrimental bone health consequences, such as osteoporosis and arthritis [6]

Lie # 2, Weight Loss: Milk DOES NOT help you to lose weight. Research conducted by Amy Jay Lanou, at the University of North Carolina, and in conjunction with the Physicians Committee on Better Medicine, concluded while evaluating evidence using forty-nine clinical trials from 1966–2007, dairy products do not help people to lose weight; in fact, it does just the opposite! Furthermore, according to Dr. Neal Barnard, dairy products are a significant factor in the contraction of obesity and diabetes! [7] & [8]

Besides obesity and bone health, recent scientific studies have also linked dairy products to increased risk of heart disease, diabetes, obesity, immune system dysfunction, Alzheimer's disease, prostate cancer, testicular cancer, ovarian and breast cancers!

Unlike organic plant milk, cow's milk contains dangerous, toxic, and disgusting ingredients, such as growth hormones, antibiotics, pesticide residues, radiation, pus, blood, along with high levels of saturated fats and cholesterol! The latest Loma Linda University and Harvard Medical peer-reviewed research document an 80% increase in the risk of breast cancer just by using the recommended daily allowance of milk under the guidelines of the USDA food pyramid!!!!! [9] & [10]

Covid-19: The mucus causing properties in dairy also play as an excellent host for Covid19, and other viruses, as well as other respiratory illnesses, such as asthma. Most of the victims of Covid-19, who succumbed to death, died of a form of pneumonia which causes congestion to form in the lungs, making it impossible to breath. As was stated in the last chapter, would you offer a glass of phlegm causing milk to a Covid-19 victim who can barely breathe? [11]

Ban Dairy, especially during a pandemic!

RECUMBENT BOVINE GROWTH HORMONE (rBGH):

The hormones injected into livestock also have deleterious effects on both cattle and human populations. These hormones simply make cattle very fat, so what do you think it does to people? In 1994, the FDA secretly approved of the genetic growth hormone referred to as rBGH. At that time, the Monsanto Corporation had already invested 500 million dollars in the creation of this hormone to increase milk production. Inevitably, this growth hormone has created suffering and disease for the cattle and to the humans who ingest it. During the time of this introduction to rBGH, dairy farmers were beginning to report that their cattle were becoming sick with a disease called mastitis.

To fight this disease required massive dosages of antibiotics. Due to this reason, 95% of dairy farmers initially refused Monsanto's pernicious protocol. However, due to the powerful influence that Monsanto had on the agricultural industry, all these ranchers finally gave in to industry pressures. This bovine growth hormone also increases the production in humans of another powerful naturally occurring growth hormone referred to as IGF-I, which has also been linked to cancer!

During this period, a scientific voice of reason was needed. Besides such prolific crusaders in this field like Jeremy Rifkin and Howard Lyman, Robert Cohen was also instrumental in the fight against Monsanto and rBGH. He publicly demonstrated his belief not only about the collusion between Monsanto and the FDA but also as to a major cover-up by the scientific community! His three years of research proved him to be alarmingly correct! In addition, his research also uncovered a connection between Monsanto, the FDA, and congressional leaders! Cohen revealed conclusive scientific evidence that laboratory animals treated with rBGH developed cancer. However, even with all this evidence, Cohen could still not convince the FDA to ban the use of this hormone. As usual, America has been sold out to the big corporations!

As was quoted from the book *Milk the Deadly Poison*:

> *"A sip of milk contains hundreds of different substances, each one having the potential to exert a powerful biological effect when taken independently of the others"*

It goes on to explain that growth hormones in milk, when combined with other hideously disgusting ingredients such as pus, blood, feces, allergenic proteins, saturated fat, pesticide residue, along with high levels of bacteria and several retroviruses (bovine leukemia, bovine tuberculosis, and bovine immune deficiency virus), are creating a myriad of negative health effects on the American public, while at the same time the dairy industry hides behind a façade of glitzy milk mustache, advertising campaigns exuding factitious health claims![12]

Antibiotics residues often show up in the milk and meat that people consume, making the human population increasingly vulnerable to more virulent strains of infectious disease-causing bacteria! These are viruses and contagions which are immune to conventional medical treatments. In addition, because of the worldwide use of antibiotics in the food chain, "new drug-resistant Super Bugs" are becoming widespread! 80% of all antibiotics manufactured in the world are fed to animals, causing human populations who consume animal products to build an immunity. This dangerous threat to

public safety is projected to kill <u>30 million</u> people by 2050! One would guess that organic plant milk (which contains none of these harmful ingredients or residues), doesn't sound too bad anymore, does it? It's a no-brainer, and it should become obvious that the people who continue to drink cow's milk are not using theirs! [13]

RADIATION:

Now, if the reader is not already thoroughly convinced, then "try this one on for size!" With the earthquake in Japan causing the most serious nuclear disaster in history, radiation from this disaster is causing worldwide fallout which is disseminated by the land, water, and air. The physical properties of milk are conducive to the imbuement of radioactive materials.

In laymen's terms, "milk sucks radiation out of the environment like a sponge!" so now on top of all the other horrible ingredients mentioned, radiation from fallout (which is conducive to cancer), is another dangerous factor that predominantly, only dairy products possess. (See chapter "The Insidious Nature of Nuclear Power") [14]

Besides catastrophic illness and egregious environmental effects, the production of animal products also represents an extremely cruel and inhumane protocol! Due to the heavy burden of industrial milk production, a dairy cow that can live up to 24 years, may only live till 4 and then get ground into hamburger meat!!! The poor animals are raped (Artificially inseminated), and thereafter, the baby calves are heartlessly stolen away at birth, and much of the time MURDERED! Anyone who claims to be vegetarian for compassionate reasons, but still uses dairy, is a Hypocrite!!! And after all this cruel and disturbing evidence, the dairy industry still has the gall to call these helpless victims "Happy Cows?" [15]

As we have learned, due to (pasteurization, hormones, antibiotics, pesticide residue, radiation and other dangerous chemicals), that dairy products are directly linked to catastrophic illnesses, including cardiovascular, osteoarthritic, cancer, digestive and immune system dysfunction. Drugstores and supermarkets in America today are stocked with drugs designed to counteract the effects of using dairy products! Some of these drugs include sinus headache pills, nasal sprays, decongestants, and antihistamines, along with stool softeners and laxatives for gas, bloating, diarrhea, and irritable bowel syndrome, all due to lactose intolerance. [16]

(For information about environmental hazards associated with dairy cattle, please refer to chapterss "Dust, Drought and Desertification, "and "Cattle, and Egregious Greenhouse Gasses"

So, let's get the facts straight: Sucking on a cow can give you cancer, heart disease, Covid-19, and other respiratory illnesses, Alzheimer's, diabetes, obesity, osteoarthritic conditions, along with antibiotic resistant strains of infectious disease! Furthermore, milk Sucks up radiation out of the environment, which even has more dangerous carcinogenic effects!

So, in conclusion, one can only come to a final prognosis that...

MILK REALLY SUCKS!

Notes:

[1]   https://www.googleadservices.com/pagead/aclk?sa=L&ai=DChcSEwiekLuSmtPzAhUx-Da 0GHSy6BMcYABADGgJwdg&ae=2&ohost=www.google.com&cid=CAESQOD23fMW2fHLDxapc Wa55pBWPRsksSCXU5kvaBPKHHjFEs2ySrWCxeXdO7cqBMybmIHgsSQYMoFagGm3Rh6rXbY &sig=A-OD64_1KPh2ILH60uNFegEpUWFJJA8hUwg&q&adurl&ved=2ahUKEwibnLSSmtPzAhW   Xrp4KHTh-bATsQ0Qx6BAgGEAE

[2]   https://www.theplantway.com/dr-benjamin-spock-vegan-diet/   [3]https://www.googleadservices.com/pagead/aclk?sa=L&ai=DChcSEwiekLuSmtPzAhUxDa 0GHSy6BMcYABADGgJwdg&ae=2&ohost=www.google.com&cid=CAESQOD23fMW2fHLDxapc Wa55pBWPRsksSCXU5kv-aBPKHHjFEs2ySrWCxeXdO7cqBMybmIHgsSQYMoFagGm3Rh6rXbY   &sig=AOD64_1KPh2ILH60uN-FegEpUWFJJA8hUwg&q&adurl&ved=2ahUKEwibnLSSmtPzAhW Xrp4KHThbATsQ0Qx6BAgGEAE

[4]   www.usdairy.com%2fabout-us%2fnational-dairy-   council/RK=2/RS=47x7JHoR_HaYDzk-PVU3tZ_GszP0-Pasteurized

[5]   https://www.foodandwine.com/news/borden-dairy-bankruptcy

[6]   Milk: Killing the Bad and the Good, http://healthyfixx.com/13/ pasteurized-milk-killing-the-bad-and-the-good.

[7]   https://medicalxpress.com/pdf128945283.pdf

[8]   https://www.googleadservices.com/pagead/aclk?sa=L&ai=DChcSEwik67L6r9PzAhV9H6 0GHTG9DPwYABAAGgJwdg&ae=2&ohost=www.google.com&cid=CAESQOD28CBJWW-   dGdW-7fTh9gnDjGIlozY-K_qROuQnQycgv9mwJfGX-   skP_HPSrJBw8PPmsDZ8bpC5T9e8eZPbZL38&sig=A-OD64_1UhffZLSSewwhroEy9cRhR0thUcg   &q&adurl&ved=2ahUKEwign6n6r9PzAhWkGDQI-HUI-BQsQ0Qx6BAgCEAE

[9]   https://switch4good.org/harvard-researchers-claim-cows-milk-is-unnecessary/

[10]   https://www.sciencedaily.com/releases/2020/02/200225101323.htm

[11]   Go Dairy Free: Is your Immune System Confused?http://www.godairyfree.org/200712042150/News/Nutrition-Headlines/Is-your-Immune-System-Confused.html

[12]   https://www.googleadservices.com/pagead/aclk?sa=L&ai=DChcSEwiG9Pubsd-PzAhVDIa0GHeizA   qQYABAAGgJwdg&ae=2&ohost=www.google.com&cid=CAESQOD2sIoX-q17WwdQJKUqcdgKK   I_vmfJ8Q5z1iKiPFqc4SHWNlINBTz1hDF9KWN6t7jTW1JTXRLosz-   UY-dvwU6rJ8&sig=AOD64_1MuU9h8mKInHME3LIcWa6JfJfR6A&q&adurl&ved=2ahUKEwintv SbsdPzAhUEJzQIHXnZB8YQ0Qx6BAgEEAE

[13]   https://www.veganfoodandliving.com/why-vegan/animal-impact/antibiotic-resis-tance-the- health-issue-thats-threatening-our-lives/

[14]   https://www.reuters.com/article/us-radiation-milk-   idUSTRE72U4GW20110331#:~:-text=WASHINGTON%20(Reuters)%20%2D%20A%20trace,U.S.%20regulators%20said%20on%20 Wednesday.

[15]   https://www.livekindly.co/joaquin-phoenix-dairy-industry-best-actor-oscars-speech/

[16]   https://plantbasednews.org/opinion/milk-is-deadly/

The Conscious Planet

CHAPTER 16

IMPLICATIONS OF GMOs

(GENETICALLY MODIFIED ORGANISMS)

All living things on this planet contain genes. The genetic makeup of living organisms forms how life works. With the science of Genetic Engineering, new genes are introduced which come from different forms of living things, but not provided by nature. E.G., genes from an animal, such as shellfish, may be combined with a fruit, such as a tomato, in order to strengthen the Fruit/vegetable's resistance to bugs, cold, or deterioration. Genes could be added to a plant from animals, viruses, bacteria, or other plants.

However, Genetically Modified Organisms are not innocuous to nature as their manufacturers claim. These new reinforced modified genes carry deleterious side effects, besides compromising nutritional value. When completely unrelated organisms can be introduced into the food chain, then this should send up red flags! Moth bacteria and fish genes are currently being engineered into plants.

The infamous Monsanto Corporation has created the herbicide resistant "Round Up Ready" soybean which contains genes from (Agro bacterium sp.), a bacteria found in a cauliflower virus, and also in petunias. These genetically modified seeds were designed to be used in conjunction with the deadly Roundup® herbicide"! Due to this seed modification, the herbicide will not affect the Genetically Modified soybean plant itself, but only the weeds around it.[1]

SENATE BILL HR 1599: MANDATORY LABELING OF GMOs

In late July of 2015, the defeat of this bill had dealt a major blow to the health of the American public! The biotech industry and Monsanto had spent 50 million dollars to overthrow it, even though 90% of U.S. citizens interviewed say that they would avoid GM products if they were clearly labeled! Monsanto had paid off the politicians so well, that the bill was defeated 71 to 27! And these are the same criminals that take our tax dollars to pay for their perks and exorbitant salaries, so that they can pass laws to help powerful companies make huge profits by completely covering up the fact that the public is being poisoned!

And people are so worried about ISIS? 71 to 27 against labeling our food to reflect that it's genetically modified along with extremely dangerous levels of Roundup® herbicide residues? We couldn't have elected a better group of terrorists! Under this new statute, a company may only voluntarily list GM ingredients. What company wants to do this if not required by law, and then they can still actually get away with saying "All Natural!"

This national legislation had eliminated local jurisdiction over GMOs, which dramatically compromised or nullified over 150 existing statutes, regulations, and ordinances, in 43 states at state and municipal levels, including GMO free zones! This law will significantly impugn the ability of organic farmers to protect their crops from contamination which could wipe out foreign and domestic markets! [2]

MISCONCEPTION: GMOS WERE DESIGNED TO PREVENT WORLD HUNGER:

Due to huge multinational corporations controlling the world's food supply, Genetic Engineering, thus could potentially drive small farmers out of business and into poverty. The motivation of the biotech industry is to control all levels of food production, starting with modified seeds, fertilizers, food processing, and water supply.

Starting in 1980, companies have been legally allowed to patent genes and living organisms, which may prevent others from growing or breeding these organisms. Therefore, if a farmer doesn't purchase GMO seeds from a bio-tech company, not only can they be fined, but as a failsafe measure these biotech companies have already developed what is referred to as "Terminator Technology," so that the seeds that the farmer saves won't grow! [3]

Monsanto would like you to believe that the weeds growing with soybeans take nutrients and water away from the soil, and therefore, by spraying Roundup®, farmers can increase yield while making it easier to

harvest. However, in a report issued by the USDA in 2006, it was stated that they "could not find positive financial impacts in either the field-level nor the whole-farm analysis."

On a similar note, the Canadian National Farmers Union (NFU), bluntly states "The claim that GM seeds make our farms more profitable is false!" As of 2010, the Canadian net farm income had significantly declined since the introduction of GMO canola. They reported that their income over last 5 years, since the inception of GMO seeds, prior to 2010, was the worst in Canada's history!

Also, in one of the most comprehensive reports on this subject ever issued, by the Union of Concerned Scientists, in 2009, entitled "Failure to Yield," they clearly demonstrated that, despite years of effort by Monsanto, the GM crops produced fewer bushels! [4]

HARVEST OF DEATH:

The use of Roundup® has been a disaster all over the world. Hundreds of thousands of farmers in India have already committed suicide by being forced to use this deadly herbicide with disastrous results! Many of them by swallowing the actual poison itself as a protest to Monsanto's nefarious protocol! This has been infamously referred to as the "Harvest of Death!" [5]

ENVIRONMENTAL HAZARDS OF GMOs:

Under Nature's paradigm why change a system that has worked for four billion years? Foreign genes are never welcomed by plant or animal, thus these powerful, yet highly unnatural elements are used to modify the structure of another living organism. The protein structure and immune system of the plant or animal is altered to accept foreign genes. There are seriously adverse effects which may present themselves! Why does man always think that he knows more than nature? Didn't Monsanto learn their lesson with the Mad Cow Disease? (Bovine Spongiform Encephalopathy). Nature never intended the bovine to ingest flesh. When you mess with Mother Nature, there will be re-percussions. (See chapter 27: "Mad Cows, Englishmen and Howard Lyman")

- ### Bio Invasion:

Like a science fiction movie, the creators of these organisms have very little control over the outcome of their creations! Could it have an adverse effect on the environment or other organisms that ingest it? In the case of Monsanto, like with Mad Cow Disease, could they be responsible for creating a deadly organism that can't be stopped?

The damage to the environment, due to widespread contamination, could become irreversible! Harmful plants, such as weeds, may build up a resistance from increased pesticide use.

Uncontrolled interbreeding of various plants spreading modified genes into wild populations may have grave long-term implications! A Genetically Engineered species, out competing wild species, in an ecosystem not able to handle them, could result in what are referred to as Super Weeds! Also, GMO plants may alter soil bacteria in ways which may impugn the integrity of the soil. [6]

- ### New Bill Would Allow Farmers to Sue Monsanto If GMO Crops Invade Their Property

Contamination from genetically engineered farming can have di-sastrous repercussions on non GMO farmers. They are in jeopardy of their crops being rejected by various export markets, which have banned GMO crops. These chemicals have also been known to lead to resistant strains of insect and weed infestation. [7]

- ### Honey Bees:

Beneficial insects, such as honeybees, are also becoming jeopardized from the increased use of Roundup® and other neonicotinoid-based herbicides, all created by Monsanto!

Colony Collapse Disorder (Where bees become disoriented and can no longer return to their hives), is affecting increasing numbers of bee populations today. This is a serious ongoing problem. 90% of wild bee populations have already been wiped out! If bees were to ever become extinct, the whole world would starve! (Refer to Endangered Insects, chapter ?) [8] & [9]

•   Could GMO Salmon Destroy the Entire Salmon Industry?

In farm raised salmon, the FDA wants to approve the use of GMOs, stating that it's "safe" for the food supply. Due to the fact that the FDA insists that the salmon spawned from genetic engineering will not be labeled; the knowledge of this should inspire a major boycott of the salmon industry!

The natural health community will launch nationwide alerts to their customers, such as Whole Food Markets. This action by the FDA is an insult and major blow to human beings on this planet who wish to protect the integrity of our food chain rather than to alienate consumers!

The FDA should form a Wholesome Food Committee, where panel members who represent non-bias scientific academia, may be able to present evidence against GMOs, and especially address the issue of labeling these unwholesome foods. The FDA should protect the consumer by banning GMOs in the first place but allowing it to be produced and go unlabeled is downright sinister!

If this despicable practice is allowed on salmon, then it could eventually lead to the production of farm animals. It is Monsanto's goal to eventually modify all seeds, plants and animals! Yikes!

Can you imagine a heartless and greed driven cattle industry putting a full cow in a box like a veal and fed antibiotics and GMO tainted grain through tubes, macabre and forebodingly similar to the way that humans were shown imprisoned in the Sci-Fi Thriller "Matrix." From salmon to livestock, and as the movie "Matrix" portends "are we next?" [10]

> "The (FDA) Food & Drug Administration is doing just the opposite of what the citizens want them to do! Instead of protecting the public, they have become one of the greatest threats to the future of our health and safety!"
>
> — (Neil M Pine)

Effects of Roundup in the Food Chain on Animal and Human Physiology:

- Tumors: Rats Fed GMO Corn Developed Large Tumors: (French Study)
- Allergies:
- Toxicity: Research indicates damage to the liver in rats who ingest GMO corn

- Altered Nutritional Value:

- Sterilization: Third generation hamsters fed GMO foods became almost completely sterile. (Russian Study)

- Non-Hodgkin's Lymphoma and Other Forms of Cancer: Humans and pets exposed to Roundup are contracting forms of blood cancer. The World Health Organization (WHO), has determined that exposure to Round Up weed killer poses a significant health risk.

- Endocrine Disruption

- Auto Immune Dysfunction

- Assault on the Mitochondria (autism, ADD, Dementia, brain fog and mental exhaustion)

- Microbiome Impairment

- Compromised Mineral Absorption and Altered Nutritional Value

- Break Down of Metabolic Pathways

Plant metabolisms become altered as genetic engineering introduces new proteins into the food chain. This leads to a greater propensity of allergic reactions, as well as toxicity, and altered nutritional value. Without warning labels, an individual with an allergy to shellfish could theoretically have a severe reaction just by eating a vegetable treated with this gene! Also, the possibility of antibiotic resistant strains of infectious diseases could exist. Antibiotic resistant genes may transfer into intestinal bacteria and exacerbate health problems. Antibiotics, which once treated existing conditions, may now become benign in its effects. Roundup® acts like an antibiotic. [11]

FRENCH RESEARCH STUDY:

In September of 2012, a French research team published conclusive evidence, in a U.S. peer reviewed journal, exposing the highly dangerous effects of GMOs and Roundup®! Thanks to Monsanto, GMOs have permeated the very fiber of our society! In America today, the majority of non-organic corn, soy, cotton, and sugar beets, are all genetically modified! In this study, rats fed GMO corn developed large tumors![12]

Russian Biologist Documents Sterility in Hamsters Fed GM Soy!

Russian biologist, Alexey V. Surov, has uncovered conclusive evidence of a connection between sterility and a dramatic rise in infant mortality rates, from hamsters fed GM soy!

> *"By the 3rd generation, most GM soy fed hamsters lost their ability to have babies!"*

> — (Alexey V. Surov) [13]

Dr. Greger on the Serious Dangers of GMOs!

In his 2014 video, Dr. Michael Greger substantiated research which documented that GMOs and their key ingredient Roundup®, are highly toxic and adversely affect cell tissue, citing one study using human placenta! He also refers to a study which examined the long-term effects of GMO crop contamination. Globally, as of 2011, Roundup Ready G.M. soybeans represented 75% of total soy production! In the preliminary stage of this study, researchers found that levels of Roundup® sprayed on soybean crops, had been exceedingly high as compared to the legal maximum allowable residue levels. At that time, the legal limit for glyphosates (The active ingredient in Roundup®), was set at only .1 milligram per K.G. During this test, it was discovered that the legal limit for herbicide residues were exceeded by up to 2000%!

Conventional non-GMO, and organic soy products, contain NONE of these dangerous toxins! And what was Monsanto's response to these extremely hazardous levels of pesticide residues? Did they decide to follow the legal guidelines under the Health and Safety code, set forth by the USDA and the FDA? The action they took next can only be described as being at the height of criminal negligence!

In the face of documented research, which exposes the serious nature of Roundup®, Monsanto, with all their power, was successfully able to circumvent all accountability, by influencing the authorities to actually raise the safety limits of their deadly herbicide residues by up to 20,000%! No type o; Twenty Thousand!

In 2004, Brazil increased its limit for Roundup® herbicide residues by 5000%, from .2 mg per K.G., to 10 mg per K.G. Even more significant, was Europe and the U. S., which in 1999, increased this toxic residue standard by a full 20,000%, from .1 mg to 20 mg per K.G.! In this way, those old readings don't look so high anymore, do they? Now on top of all this, making it legal to use 200 times more Roundup® would also dramatically increase sales!

> "The acceptance level of glyphosates in food and feed, i.e. the maximum residue level (MLR) has been increased by authorities in countries where Roundup Ready GM crops are produced or where such commodities are imported."

In addition, Dr. Greger points out, that it's not so much the active ingredient in Roundup® (Glyphosates), that makes it so toxic, but rather the chemical Surfactants and Adjuvants which comprise this deadly cocktail, also known as Roundup®! It was found that this herbicide was 125 times more toxic than the active ingredient (Glyphosate), by itself, but when Monsanto refers to their research, they only tested this active ingredient by itself to justify their safety findings! [14]

ROUNDUP® INGREDIENTS:

> Here is a List of What You're Actually Consuming in Your Food if You Eat Major Brand Name or Fast Food Products, or Eat Non Organic Soy, Corn, or Processed Sweeteners!

The active ingredient is Isopropylamine salt of glyphosate, along with a variety of chemical adjuvants used as a catalyst to activate the glyphosates. Besides the glyphosates, this toxic formulation also includes Ammonium Sulfate, Benzisothiazolone, 5-chloro-2-methyl 3(2H)- 2-Propynyl Butylcarbarnate, Isobutane, Isopropylamine, Light Aromatic Petroleum Distilate, Methyl P-Hydroxy-Benzoate, Methyl Pyrro-Lidinone, Pelargonic Acid, Polyethoxylated Tallowamine or Alkylamine (POEA), Potassium Hydroxide, Propylene Glycol, Sodium Sulfate, Sodium Benzoate, Sodium Salt of O-Phenylphenol, and sorbic acid.

Why Were These Products Not Tested for Safety?

There was never any safety study conducted after the FDA declared, in 1992, that GMO foods are "Substantially Equivalent" to regular foods, and therefore, were never regulated for testing. In 1999, a major lawsuit was filed against the FDA, suggesting that the FDA's own scientist had covered up evidence which concluded that GMOs present unique safety hazards, and at that time, it was recommended that rigorous case by case safety testing be initiated. These safety warnings by the FDA's top scientists were all ignored! [15]

Documentaries by Jeffery Smith: Executive Director of the Institute for Responsible Technology:

According to the latest research presented by Jeffery Smith, author of *The Seeds of Deception* and producer of documentary films' *Genetic Roulette* and *Secret Ingredients*, it states that besides cancer, Roundup® in the food chain is also causing small holes to perforate human stomach lining, in what is referred to as leaky gut syndrome, along with a myriad of other serious medical conditions, including autism, crones, asthma, diabetes, obesity and many more! He also documents serious environmental issues associated with the use of this pernicious weed killer.

- "Genetic Roulette, the Gamble of Our lives!" [16]
- "Secret Ingredients" SecretIngredientsMovie.com [17]

> *"Genetically modified foods disrupt the major fundamental components that drive our health!"*
>
> — (Jeffery Smith)

Dr. Michelle Perro, named one of the top pediatricians for 5 years running, and author of the book *What's Making Our Children Sick,* has documented the miraculous change in sick children who have eliminated GM ingredients from their diets.

While watching Mr. Smith's documentary, *Genetic Roulette,* she realized that "this is what's making all these kids sick!" According to Dr. Perro, when she changed the diet of her young patients "the results were remarkable!" From cancer, asthma to diabetes and even autism; totaling over 30 serious conditions, she witnessed dramatic improvements across the board with hundreds of testimonials! [18]

2015: The First Major Class Action Law Suit Against Monsanto!

In 2015, Monsanto was sued in the U.S. District Court in New York. Earlier that year, with mounting evidence against Monsanto, the (WHO)'s International Agency for Research on Cancer (IARC), issued a serious health warning, linking the deadly weed killer Roundup® to Non-Hodgkin's Lymphoma and Leukemia.

Furthermore, research indicates that exposure to Roundup® compromises the human digestive system, lowers metabolism and other critical immune system functions.

According to the lawsuit, not only does this weed killer lead to resistant weeds, but it also causes resistant strains of stomach bacteria, leaving humans highly susceptible to opportunistic pathogens, indigestion and other health problems! [19]

Major Lawsuit Settlements Against Monsanto Originally for Over 2 Billion $ With only 4 People and Almost 19,000 More Victims Waiting in Line!

2019 was a banner year for plaintiffs in major litigations against the chemical giant Monsanto, who was purchased by Bayer Pharmaceuticals in early 2019. The first case (Johnson vs Monsanto), was originally settled for 289 million, but was later reduced to 78.5 million dollars. The second case (Hardeman vs Monsanto), was settled for 80 million dollars. However, in the most recent case (Pilliod vs Monsanto), the original verdict had stood at a whopping, 2.055 Billion dollars! However, as of 7/25/2019, California Supreme Court Justice, William Smith, slashed this award by more than 90%, down to only 86.7 million! It has been speculated on by Monsanto's legal defense team, that the other 18,400 litigants so far, will be compensated through a class action for between eight to ten billion dollars. Due to the ever-mounting list of victims, many people feel this figure to be grossly inadequate. [20]

In 2019, in light of these litigations, sensing people's fears, even major brand names like Dorito's, Post, and General Mills, are starting to come out with Non-GMO or organic labeling, for their chips, cereals and crackers.

The implications of these judgments will be devastating for Monsanto and their parent company, Bayer Pharmaceuticals, whom have consequently, as of October, 2019, had lost approximately 40 billion dollars in market capitalization! The end result of these class action findings had laid the groundwork for multi-billion-dollar litigations against Bayer/Monsanto. On 6/24/2020, the court finally ordered Monsanto to pay its victims almost 11 billion dollars! This represents almost 19 thousand plaintiffs. [21]

But for them, may this author attempt to play the world's smallest violin!

HERE'S WHY:

Since 1981, Monsanto had been hiding pro- prietary trade secrets which indicated serious health risks from exposure to Roundup®. Solid evidence proves, that for years, Monsanto was manipulating data and bribing EPA officials to cover up the danger of their deadly herbicide and in 2016, the WHO (World Health Org.), final- ly concluded that this insidious weed killer was highly carcinogenic and had been directly linked to several types of cancer and organ damage.

Not only did agricultural workers become the victims, but also landscap- ers, gardeners, and even children and animals who may have played in/or occupied the areas where these gardeners and/or landscapers were work- ing! [22]

Furthermore, our food chain has become dramatically impugned by us- ing this pernicious herbicide. The health and safety of the public have been seriously endangered due to the production of Genetically Modified Crops which contain dangerous levels of this herbicide Roundup®!

Ammonia, is a safe, extremely cheap, plentiful and naturally occurring nontoxic element, which can also be used to kill weeds. Since this preceden- tial court decision against Monsanto, even major food brands like Dorito's, Post and General Mills, are starting to come out with Non – GMO or organic labeling for their chips and cereals, sensing the public's legitimate concern! In 2017, Roundup® was directly linked to 70,000 cases of cancer annually!

(Please refer to link to the article below)

A dangerous law was passed during the Trump Administration near the end of his term, eliminating almost all oversight of genetically engineered plants and animals. This means that chemical companies get to make their own decision, whether or not, their products need to be regulated! This au- tomatically gives them a green light to launch untested GE foods to market with no oversight! This absolutely represents a major "conflict of interest!"

Organizations like The Center for Food Safety, are challenging this rule in court. However, these corporations are ruthless! They will do everything they legally can to drag the case out as long as possible, so they can get away with their nefarious scheme! There is already strong evidence that this new rule violates numerous agricultural and environmental statutes.

With more awareness and support from the public, then swift abatement of dangerous GMO practices should be achieved! After all the major lawsuits already suffered by Bayer/Monsanto, people just don't trust GE foods anymore! [23]

2019: Petition to Remove Monsanto's toxic Roundup® from our Children's Food

In 2019, 10 food manufacturers and retailers all signed the EWG's Environmental Working Group's petition to dramatically limit the amount of glyphosate residues found on oats and also to prohibit the herbicide's use as a drying agent.

With work being conducted by people like Pediatrician, Michelle Perro, and by documentary film maker, Jeffery Smith, the serious nature of this herbicide has come to light. The public is demanding healthy and safe nutrition for their children! [24]

How Can I Protect My Family From Genetically Modified Foods? Tips for Avoiding GMO's:

1. Insist on organic or local and farm raised produce, or non-GMO labeling (non-pesticide variety if not certified organic). Organic standards prohibit the use of GMO's. Look carefully at organic labeling; there are 3 types. Some may be deceptive. The best kind states 100% organic. This means that all ingredients are certified organic; no GMO's can be legally used in the production of this product. The next best rating is organic or certified organic, which means that at least 95% of all listed ingredients are organic. Also, the remaining 5% cannot contain GMO's either. When labeled, "Made With Organic Ingredients;" this technically means that only 70% of the ingredients need to be organic. However, according to FDA standards, the remaining 30%, (even though non- organic), also must not contain any GMO's. But, if the term organic is only in the listed ingredients, and not on the label or package, then no required % for organic ingredients is required, and therefore, any non-organic items may contain GMO's.

2. Look specifically for non-GMO labeling. Many conscientious health-conscious corporations voluntarily label their products.

3. Avoid suspect foods; the major 7 GMO crops to be aware of are Soy, Corn, Cotton Seed, Canola, Hawaiian Papaya, Zucchini, and yellow squash. Carefully study the variety of non-GMO counterparts. By physical observation, you can see the difference. Agricultural products such as seedless watermelons, pear-apple combos, and tangelos, are all products of natural breeding and are not Genetically Engineered. In the U.S., people mainly need to avoid processed foods made with soy, corn, or wheat. It has been estimated that 90% of all processed foods contain some form of corn or soy products.

4. Look for non-GMO shopping guides. At truefood-now.org, they offer non-GMO brand choices for the informed consumer. Also offered is a book by Andrew Kimbrall, *Your Right To Know*. This book is available at Seeds of deception.com. For healthier non- GMO eating, also check out Responsibletechnology.org

5. There are other hidden GMOs to look out for. All artificial sweeteners use GMOs. These products are everywhere and in everything. More than 6000 products contain GMOs; soft drinks, gum, candy, desserts, snack mixes, tabletop sweeteners, vitamins, cough drops and other over the counter remedies, and many prescription pharmaceuticals.

Also, animal products; meat, dairy, eggs, and farm raised fish, are usually originated from animals being fed GMOs in their feed. The only way to circumvent the risk of GMO exposure is buy organic meat and dairy products. In addition, honey and bee pollen could have been gathered from GMO plants. In general, avoid animal products and processed foods, as they can contain a variety of GMO ingredients. Better yet, Go Organic Macrobiotic Vegan! [25]

HOW DO WE STOP THIS?

- Make an Effort to buy only non-GMO products

- Buy Organic When Ever Possible

- Let Everybody Know How You Feel About GMO's (Super Markets, Restaurants, etc...)

- Call All Food Manufacturers and Let Them Know How You Feel About GMO's Effects on the Food Chain

• Call Your State Representative to Voice Your Opinion (Ask for Strict Testing Regulations)

• Vote For Candidates Who Support This View of Stopping World Hunger and Banning GMO's

• Make a Statement by Practicing an Organic Vegan, Macrobiotic Diet, Free From harmful GMO's.

In a survey conducted by Novartis Pharmaceuticals, they discovered that 93% of Americans support legislation that genetically altered foods should be labeled as such! However, our government currently requires no warning labels and in fact, is supporting the Bio-tech industry's right to refuse to provide warning labels, despite the fact that food producers want to cooperate with consumer sentiment!

The smartest solution to this problem would be to conduct test trails on people who only have a vested interest in this insidious technology. That's right, if they want to make money on it then they should be their own Ginny pigs! After 5 years, if they are still alive, and haven't grown an extra head, then maybe we could let the public try it. But still, not without a warning label! [26]

Monsanto: The Shocking History behind the World's Most Insidious Corporation!

This is the same benevolent corporation that has embellished us with such wonderful and wholesome products over the years like Napalm, Agent Orange, DDT, Roundup®, recumbent Bovine Growth Hormones (rBGH), and Aspartame!

How Monsanto Was Able to Achieve a Wanton Disregard for all Public Safety Worldwide: The origins of the infamous Monsanto corporation came straight from the bowels of Nazi Germany! The I.G. Auschwitz Chemical Company, founded in 1941, was designed to be the largest chemical factory in Eastern Europe! Forced slave labor from Nazi concentration camps was utilized, where workers were exposed to highly toxic chemicals and unsafe working conditions.

The I. G. Farben corp., which at that time, was comprised of the current BASF corp., the Bayer corp., and the Hoechst corp. (Now known as Aventis), owned I.G. Auschwitz. Frits ter Meer, was managing director of operations for

I.G. Auschwitz. Under his direction, the use of sodium fluoride was utilized in the drinking water of concentration camp victims, in order to control and reduce their populations. The side effects on these victims included toxic effects to the kidney and liver, brain damage, Alzheimer's, breakdown of the central nervous system, and chromosomal damage. Frits ter Meer was ultimately convicted of a war crime at Nuremburg, but due to his powerful connections, he was actually able to get his sentence commuted. He subsequently became one of the original founders of Codex Alementarius Commission in 1963.

Kiss your VITAMINS Goodbye

The United Nations and CODEX ALIMENTARIUS have a plan for you

These were Monsanto's "Good Old Boys," the notorious Codex Alementarius Commission! Even the name sounds austere and foreboding. The roots of this organization are shocking, stemming from a precept of global eugenics and world domination!

> "The Codex Alementarius Commission was created in 1963 by the Food and Agriculture Org. (FAO), and the World Health Org. (WHO) to develop food standards, guidelines and related texts such as codes of practice under the joint FAO/WHO, Food Standards Programme. The main purposes of this Programme are protecting the health of the consumers and ensuring fair trade practices in the food trade, and promoting coordination of all food standards work undertaken by international governmental and non- governmental organizations"
>
> ---- (Codex Alimentarius Commission)

However, in essence, this organization merely serves as a front for international biotech, drug, and chemical companies such as Monsanto. Obviously, rather than to serve in the best public interest, their ultimate goal is to help mitigate safety regulations levied against gross polluting corporations and to disseminate surreptitious information to the public regarding food safety and nutritional supplements.

> "Their actual goal is to outlaw health products and information on vitamins and dietary supplements, except those under their direct control. These regulations would supersede United States domestic laws without the American people's voice or vote in the matter!"

"Instead of focusing on food safety, Codex is using its power to promote worldwide restrictions on vitamins and food supplements, severely limiting their availability and dosages."

THE HISTORY BEHIND THE CODEX ALEMENTARIUS COMMISSION:

The I.G. Auschwitz Chemical Company, founded in 1941, was designed to be the largest chemical factory in Eastern Europe! Forced slave labor from Nazi concentration camps was utilized, where workers were exposed to highly toxic chemicals and unsafe working conditions. The I. G. Farben corp., which at that time, was comprised of the current BASF corp., the Bayer corp., and the Hoechst corp. (Now known as Aventis), owned I.G. Auschwitz. Frits ter Meer, was managing director of operations for I. G. Auschwitz. Un-

TER MEER, Fritz

der his direction, the use of sodium fluoride was utilized in the drinking water of concentration camp victims, in order to control and reduce their populations. The side effects on these victims included toxic effects to the kidney and liver, brain damage, Alzheimer's, break down of the central nervous system, and chromosomal damage. Frits ter Meer was ultimately convicted of a war crime at Nuremburg, but due to his powerful connections, he was actually able to get his sentence commuted. He subsequently became one of the original founders of Codex Alementarius Commission in 1963.

Therefore, the same original Nazi organization, which poisoned the Jews with fluoride, has also become instrumental in setting worldwide policy standards regarding fluoride in drinking water and approving of the ridiculously high levels of toxic herbicides found on today's GMO crops! (Refer to chapter on Pollution)

The Codex Alementarius Commission, through the (WHO), World Health Org., recommended fluoridation all drinking water, worldwide!

"Many countries are currently undergoing nutrition transition do not have adequate exposure to fluoride"

— (Codex Alementarius Commission)

However, while expert committees from the (WHO) recommended that the worldwide level of fluoride exposure, be set at .5 to 1 mg/L, Germany on the other hand, uses absolutely NO fluoride in their drinking water! So, is it any coincidence that one of the main founders of the Codex Alementarius Commission, just happened to be in charge of the fluoridation program at Auschwitz? Obviously, they know something WE don't! [28]

How is Codex Connected to Monsanto?

The Aventis corp. (formally the I.G. Farben corp.), is currently partners with Monsanto and Cargill corp., along with their original partners from the "Old Country;" the BASF corp., and Bayer Pharmaceuticals, respectively. They are all involved with pseudo "Crop Science." In 1998, Monsanto teamed up with Cargill taking over their global seed cartel. Later, Bayer partnered with Monsanto to work on herbicide tolerance, and BASF has worked with Monsanto to increase crop yield. [29]

Notes:

[1]   https://www.cnn.com/2019/02/14/health/us-glyphosate-cancer-study-scli-intl/index.html

[2]   http://naturalsociety.com/senate-shoots-down-gmo-labeling-bill/

[3]   https://cases.open.ubc.ca/monsanto-and-terminator-seeds/

[4]   https://www.ucsusa.org/resources/failure-yield-evaluating-performance-genetically-engi-neered- crops

[5]   http://www.seattleorganicrestaurants.com/vegan-whole-foods/indian-farmers-commit-ting- suicide-monsanto-gm-crops/

[6]   https://www.frontiersin.org/articles/10.3389/fbioe.2019.00454/full   information@earth-save.org

[7]   https://www.naturalnews.com/2017-03-27-new-bill-would-allow-farmers-to-sue-monsan-to-if- gmo-crops-invade-their-

[8]   https://www.sciencemag.org/news/2018/04/european-union-expands-ban-three-neonic-otinoid- pesticides

[9]   https://www.npr.org/2018/09/25/651618685/study-roundup-weed-killer-could-be-linked-to- widespread-bee-deaths

[10]   http://www.naturalnews.com/029957_genetically_modified_salmon.html

[11]   http://www.responsibletechnology.org/posts/category/blog-posts/page/18/

[12]   www.motherjones.com/tom-philpott/2012/09/gmo-corn-rat-tumor

[13]   http://veganskeptic.blogspot.com/2011/10/alexey-surov-and-gm-soy-recurrent-tale.html

[14]   https://drgreger.org/

[15]   https://usrtk.org/the-fda-does-not-test-whether-gmos-are-safe/

[16]   http://geneticroulettemovie.com/

[17]   https://secretingredientsmovie.com/

[18]   https://www.chelseagreen.com/product/whats-making-our-children-sick/

[19]   https://www.aboutlawsuits.com/roundup-class-action-lawsuit-85070/

[20]   https://www.cbsnews.com/amp/news/jury-awards-couple-2billion-monsanto-round-up-weed-killer- cancer-lawsuit-trial-today-2019-05-13/

[21]   https://www.nytimes.com/2020/06/24/business/roundup-settlement-lawsuits.html

[22]   https://www.bloomberg.com/news/articles/2017-03-14/monsanto-accused-of-ghost-writing-papers-on- roundup-cancer-risk

[23]   http://www.centerforfoodsafety.org/

[24]   https://www.ewg.org/news-insights/news-release/10-food-companies-join-ewg-glyphosate-petition

[25]   http://www.mnn.com/.../5-ways-to-protect-your-family-from-gmos

[26]   http://r.search.yahoo.com/_ylt=A0SO8y2XqeVVilgA5IVXNyoA;_ylu=X3oDMTEyNmVqb2t-jBGNvbG8DZ3E   xBHBvcwMxBHZ0aWQDQjAzNjJfMQRzZWMDc3I-/RV=2/RE=1441143320/RO=10/RU=http%3a%2f%2fwww.smallfootprintfamily.com%2fhow-to-help-stop-   monsanto/RK=0/RS=ld-ou_PjGMHhrSliVkzokRW4x9L0-

[27]   http://www.natural-health-information-centre.com/codex-alimentarius.html

[28]   http://saladin-avoiceinthewilderness.blogspot.com/2010/11/codex-fluoride-auschwitz-monsanto.html

[29]   https://theecologist.org/2017/sep/14/opaque-world-codex-alimentarius-and-monsanto-toxic-relations

STOP MONSANTO!

and become a part of...

The Conscious Planet

Copyright © Neil M Pine, 2020

"Speak To The Earth
And It Shall
Teach Thee"

— (JOB 1:18)

CHAPTER 17

ORGANIC VS. CONVENTIONAL PRODUCE

In this chapter, we will try to explore and address many issues and questions regarding organic foods. The labeling, the costs, along with the ecological and health benefits, are all discussed. In making a commitment toward a healthier lifestyle, organic is absolutely a major step in the right direction. In addition to obvious health benefits, there are also factors of food safety and environmental sustainability.

In order to meet USDA certified produce standards, all agriculture must be grown in an organic medium for at least 3 years (Free from (GMO)'s, Genetically Modified Organisms, Synthetic Pesticides, or Petroleum, and/or Sewage Sludge based fertilizers. GMO foods, (Plant or Animal), contain genetically altered DNA, which should never qualify for organic labeling.

(See chapter "Implications of GMO s") [1]

Also certified organic meat and dairy products must come from livestock that were raised outdoors and given all organic feed. Under USDA guidelines, these animals must not be given antibiotics, growth hormones, or any animal byproducts. In addition, under Federal guide lines, organic food should never be irradiated. Organic produce has also been known to contain higher levels of vitamins, minerals, antioxidants, and flavonoids. Organic spinach can contain up to 200% more iron than its non-organic counterpart. An organic peach picked ripe off the tree may contain up to 12 times the flavonoid count, (Flavor), over non organic peaches which may have been picked prematurely, (when they were still hard and unripe), to avoid bruising, and must now ripen in the box or on the store shelf. These conventionally grown peaches, many times, look cosmetically appealing, but lack the satisfying flavor of a real ripe peach.

Another health benefit reported by people who switch to organic, was that people with food allergies, or who are sensitive to chemicals and preservatives, found that their conditions dissipated or simply went away, when they primarily only ate organic foods. Organic produce should never contain chemicals such as fungicides, herbicides, or insecticides. These dangerous chemicals are standard practice in non organic produce, and may leave residues which cannot be washed off! [2]

These toxic chemicals are especially dangerous for new born infants and developing fetuses, which have less developed brains and immune systems. Early exposure to these toxic chemicals has been known to cause motor dysfunction, behavioral disorders, and developmental impediment. Pregnant women are also very vulnerable to these toxins, as their breast milk may become tainted by exposure.

However, most people over the years have developed a culmination of these chemicals, referred to as "Body Burden," which has been proven to lead to cancer, birth defects, headaches, and a deficient immune system.

It's obvious that organic produce is usually fresher, tastes better, and is healthier for you. But did you know that Organic farming represents environmental awareness and conservation principles? Organic farming is demonstratively better for the environment. This practice reduces air, water, and land pollution. It conserves water, reduces soil erosion, while increasing fertility and using less energy. It is also good for birds and other small animals, as pesticides in their habitat may affect reproduction or even kill them. Organic produce is also much safer to the humans who harvest it. However, just because a product is labeled organic, some people may neglect to wash their produce. Coliform bacteria, which is a contingent factor in manure used for growing organic produce, is a dangerous pathogen; so always thoroughly wash your produce, organic or not! [3]

Negative Environmental Aspects of Non-Organic Farming:

Conventional produce farmers, rather than to replenish or enrich the soil through organic composting, they use chemical fertilizers, which degrade and erode the soil, leading to unhealthy plant species which are devoid of nutrients. The quality of the soil is a direct reflection of the health, and the health benefits, of the plant. With no regeneration of nutrients in the soil, but rather a toxic concoction of manmade chemicals, what is left is a quagmire of sick and contaminated soil. These unhealthy soils quickly erode, thus poisoning underground water tables, rivers, or lakes, which become contaminated by runoff. The American public must pay the price of this environmental time bomb! Already there are towns in the Midwest where the levels of pesticides in the drinking water are unfit for human consumption, not to mention the increased rates of cancer in these areas.

The irony is that organic farmers should be rewarded for practicing safe ecological farming, but instead they are punished by the big Agra business lobbyists, who do their best to keep the price of organic produce from becoming competitive through government subsidies. However, they are fighting a losing battle, because the only way they can win will be by the death of us all!

"There are at least 107 active pesticide ingredients that are known to be carcinogenic (cancer causing)"

The way we can win now is to inform the public of this farce and conspiracy. When people think of fresh produce, they think of good health, not of toxins and poisons. Not only can commercial produce be dangerous, but statistically speaking, organic produce can contain up to 3 times the vitamin and mineral content, besides not containing harmful residues.

A recent study conducted by The University of California at Davis, concluded that organic farming is just as productive as commercial farming, but does not destroy the soil, pollute ground water, or poison farm workers. Judging by the increased nutritional benefits, without the side effects of pesticides, Organic produce seems to be a real bargain. [4]

The Federal government reports 20 million cases of food poisoning every year. Much of this is due to agriculture related illness. If you want to reduce the cost of health care in America, increase the amount of nutrition you receive, and save our environment, then buy Organic produce!

It is the ignorance of the American public which lets big agro-business thrive. But people are becoming more aware, and it is this awareness which will eventually lead to lower organic produce costs, and in turn will create more of a demand, for what this author believes that there are no alternatives for!

As a rule, thick skinned non-organic produce, like coconuts or avocado, are generally safer than other conventional produce which may have sensitive membranes or delicate leaves which are directly exposed to toxins. [5]

Here is a list of some examples of non-organic produce which are considered to be relatively safe:

These are the types of produce which statistically have the lowest levels of pesticide residues

1. Avocados

2. Coconuts

3. Asparagus

4. Broccoli

5. Cabbage

6. Eggplant

7. Kiwi

8. Watermelon

9. Mango

10. Onion

11. Papaya

12. Pineapple

13. Sweet Potatoes

14. Peas

This list identifies fruits and vegetables which should only be organic.

The conventional counterparts of this group can be highly toxic [4]

1. Apples

2. Bell Peppers

3. Carrots

4. Celery

5. Cherries

6. Grapes

7. Kale

8. Lettuce

9. Nectarines

10. Peaches

11. Pears

12. Strawberries

13. Blueberries [6]

VEGANIC FARMING: BEYOND ORGANIC:

Do we really need manure, bone meal, blood meal and other agricultural products that harm or exploit Animals to grow healthy produce? There is a new agricultural revolution taking the vegan community by storm! Veganic Farming utilizes only organic Vegan sources of nutrients, and represents the cleanest, most compassionate and sustainable form of agricultural production. [7]

Notes:

[1]   lifegoddess.com/2010/09/01/a-quest-for-nutrition-part-3-organic-vs-conventional-food/

[2]   http://www.croplifefoundation.org/Documents/ResearchBriefs/hand-weeding-report-to-McMillan.pdf

[3]   Gary Null's Ultimate Anti-Aging Program, Null, (07:p 265)

[4]   "Documenting Food" by Catherine Butler: Producer of "The Future of Food," A documentary explaining the complex web of market and political forces that are changing what we eat, as huge multinational corporations seek to control the world's food system. The film also explores alternatives to large-scale industrial agriculture, placing organic and sustainable as real solutions to the farming crisis today! www.greencenturyinstitute.org/salon.html.

[5]   http://www.the good human.com/2008/10/22/which-fruits-vegetables-to-buy- organic/2009/08/11/to-organic-or-not-to-organic-its-an-obviou-choice/2009/07/12/pesticides-   fertilizers-hrbacides-and-water-pollution

[6]   http://www.pbs.org/wnet/need-to-know/health/the-dirty-dozen-and-clean-15-of-produce/616/

[7]   gentleworld.org/beginners-guide-to-veganic-gardening

Become Sustainable,
Become Organic,
Become a Part of …

The Conscious Planet

CHAPTER 18

ETHANOL
Unviable Government Subsidized Tax Scam

"*T*he Unraveling of the Ethanol Scam,*" by Robert Bryce [1], was a report featured in the book/movie: *Food Inc.*, which exposes ethanol and bio-fuel manufacturing as polluting and unprofitable, citing a trend toward many bankruptcies in the industry. While at the same time bureaucrats receive billions of dollars in government tax subsidies. Some experts believe that the rate of bankruptcy could reach 20%. In addition, what is most disturbing is that producing ethanol drives up the price of food! With such factors as rising energy prices and grain demand in other countries "burning food for fuel," merely exacerbates this critical scenario. [1] Environmentalist, Robert Bryce has uncovered 15 research reports which overwhelmingly indicate the unprofitable nature of producing ethanol and biofuels. One of the most significant of which, comes from David Pimentel, renowned Professor of Ecology at Cornell University. He has concluded from over 2 decades of research, that ethanol has been a contingent factor towards inflated pricing in beef, chicken, pork, eggs, breads, cereals, and milk. According to Pimentel's study *"Growing crops for bio-fuel not only ignores the need to replace natural resource consumption, but exacerbates the problem of malnourishment worldwide by turning food grain into bio-fuel."* After all his efforts, Mr. Pimentel expressed his stern consternation and disbelief toward congress's continued support of the grain ethanol industry.

Besides other factors which affect the price of food, such as high energy prices and increased global demand, ethanol also represents a significant factor in the price of food production. As Pimentel states, *"It's abundantly obvious that the corn ethanol industry has had an effect on food prices."* In 2008 alone, more than a billion bushels of corn, (one third of all the U.S. corn production), was processed for fuel. The Bureau of Labor Statistics reported in 2008, a 6% increase in the price of food, and this coming after a 4.8%

increase in 2007. At that time, the U.S. was in a recession with record unemployment and a floundering world economy.

In a research study by Tom Elam, Indiana- based agricultural economist and long-term anti- ethanol proponent, he articulately explains the current state of the ethanol industry, referring to the shutdown of 32 distilleries thus far. *"Thus, about 16.1% of all ethanol capacity in the U.S. has been idle due to the high corn costs- which are in part, a reflection the ethanol industry's own demand for grain."* [2]

In 2021, NASA scientists had reported that the state's water reservoirs could run dry in just one year's time! California could face rationing! In a joint study by The Pacific Institute and the University of California at Berkeley, they analyzed the actual amounts of water used to produce biofuel. What they found was very disturbing! The enormous water toll that livestock place upon the environment, combined with this increased production of biofuels, has led to a four-fold increase in water demand in the state over the last two decades! In fact, it takes 1,700 gallons of water to produce just one gallon of biofuel, (corn ethanol.)

With the worst drought in history, such a profligate use of water defies all logical explanation? However, the bumbling bureaucrats in Washington have mandated this policy! The Renewable Fuel Standard, demands that billions of gallons of biofuels, such as corn ethanol, are produced each year with no regard for the enormous strain this puts on California's water resources! [3]

Another factor little discussed by ethanol supporters is the actual statistics which seem to indicate an increased negative effect on air quality and yet the EPA also chooses to ignore this data! William Becker, Executive Director of the National Association of Clean Air Agencies, representing the pollution control authorities from 49 states and 165 metro areas all over the U.S., clearly states: *"More ethanol means more air pollution. Period!"*

Milton Friedman, Nobel Prize – winning economist, while discussing the exorbitant cost of ethanol production, publicly declared, *"There's no such thing as a free lunch."* Based on this research and other similar reports, Nicholas Hollis, president of the Agribusiness Council, (an agricultural trade group), has concluded, while taking into consideration the high costs of production along with the fact that this industry has always been contingent upon federal handouts, that *"Ethanol is the largest scam in our nation's history."*

So what about bio-fuels which use only renewable non-food materials?

8/31/09, *The Wall Street Journal* came out with some sobering news regarding the bio-fuel industry. *"The bio-fuels revolution that promised to reduce America's dependence on foreign oil is fizzling out."* Producers of **bio-fuel** which use renewable nonfood materials, such as sugar cane and corn stalks, are finding it impossible to meet production quotas as well as attract investment capital. *"Two Thirds of U.S. biodiesel production capacity now sits unused, reports the National Biodiesel Board."* *[4]*

Notes:

[1]   www.counterpunch.org/2007/03/02/the-ethanol-scam

[2]   http://www.news.cornell.edu/stories/July05/ethanol

[3]   https://smarterfuelfuture.org/blog/details/wasting-water-on-ethanol-amid-california-drought/?fbclid=IwAR1tC3OmLx36pkNJuDdvM9VYke_IDKUuwLat4MCI-5isWI7b1zLPwf6YBrc

[4]   blogs.wsj.com/.../04/is-corn-ethanol-cleaner-than-crude-oil

The Conscious Planet

CHAPTER 19

HYDROGEN

Not Just More Burecratic Hot Air

While the viability of hydrogen may still be years away, the future potential of this alternative energy may someday encompass a worldwide domination of the energy markets. Right now only 10-12 huge multinational oil companies currently control the flow of oil and the fate of the world's economy along with it.

We must break free from the bonds of their collusion. Technology will set us free from the deep carbon foot prints and well financed, (fossil fuel related), terrorist regimes. This new freedom represents energy independence. The U.S. and especially California, has the greatest technological knowhow with the most prestigious high-tech companies on earth! As Gov. Schwarzenegger said that, *"California will lead the nation and the world in green technology."* Hydrogen, just by itself, shows vast potential for future economic predominance and will someday drive a global economic presence which represents a plentiful and innocuous source of energy which can be manufactured anywhere.

> *"Anthropologist say that the amount of energy consumed per capita in a society is a good measure of its relative state of advance. Western society, over the last 200 years, has consumed more energy per capita than all other societies throughout all of recorded history put together."*
>
> — Jeremy Rifkin [1]

The successful advancement of hydrogen energy will require mass industrial globalization of our current corporate transportation infrastructure. Japan, Bavaria, and Detroit must all be retooled and revamped. In America, this ubiquitous industrialization will represent new indigenous entrepreneurial opportunities into our capital markets. Vast fortunes will be made, but not at the expense of the environment or the lives of young American soldiers. This will represent a new era in time, one of peace and prosperity.

A new guilt free economic prosperity will pervade throughout, therefore, endowing the American public with a new sense of pride.

DRAW BACKS OF HYDROGEN FUEL CELL TECHNOLOGY:

However, like with any new technology, it does not come without its skeptics. U.S. Secretary, Steven Chu, (Former Director of the Lawrence Berkley National Laboratory), and winner of the 1997 Nobel Prize in physics, recently commented about hydrogen technology in a recent interview with MIT Publication, Technology Review on 5/14/09. To put it bluntly, he states, "There's No Future in Hydrogen Cars". Experts have estimated that the viable production of hydrogen fuel cell technology is not supposed to take place until 2025. The dynamics of hydrogen fuel technology has not yet been perfected. For automobile transportation, many hazards still present themselves. Also, storing hydrogen in fuel tanks can potentially be disastrous due to the explosive nature of the hydrogen gas, and even the filling station itself, could present a real danger.

Another negative factor involving the production of hydrogen is that it takes natural gas (a petroleum by-product) to produce it. Ninety five percent of all hydrogen today is produced from natural gas. Another negative factor involving the production of hydrogen, is that it takes natural gas, (A petroleum by product), to produce it. 95% of all hydrogen today is produced from natural gas.

Major corporations, such as DuPont and British Petroleum, have completely abandoned their interest in hydrogen technology. Experts in the automotive industry believe that hydrogen fuel cell vehicles won't be mass marketable until 2025. [2]

However, as of late 2019, Honda Corporation, has actually revolutionized the fuel cell car market! They offer these vehicles for a payment of only $350.00 a month, but what makes this such a fantastic bargain, is that they will give you free fuel for 3 years! This alternative fuel could someday revolutionize the face of the earth! Hydrogen shows vast potential for future economic predominance and may someday drive a global economic presence which represents a plentiful and innocuous source of energy which can be manufactured anywhere, therefore defusing the fossil fuel dependence factor. This will also mitigate terrorist factions who rely on proceeds from the sale of foreign oil. Hydrogen may someday provide everyone on earth with a reliable, clean, safe, and sustainable source of energy ... where we could once again proceed with Evolution rather than Armageddon! [3]

Notes:

[1]   https://www.foet.org/books/the-hydrogen-economy/
[2]   www.foet.org/books/hydrogen-economy.html
[3]   https://automobiles.honda.com/clarity-fuel-cell

The Conscious Planet

"We Cannot Solve
Our Problems
With the Same Thinking
We Used When
We Created Them"

— (Albert Einstein)

CHAPTER 20

THE INSIDIOUS NATURE OF NUCLEAR POWER

A Wolf in Sheep's Clothing

The United States and other predatory nations of the world want you to believe that nuclear power is a tame and safe source of alternative energy—DON'T YOU BELIEVE IT!

The abatement of global warming should never be contingent upon the proliferation of nuclear power! Veteran neutron scientists Radhakrishnan of India, [1] and the late John Gofman of the United States, [2] would both vehemently remonstrate in favor of this position!

For decades, atomic scientists have speculated through sheer mathematical odds, as to the inevitable possibilities of cataclysmic nuclear disasters. Nuclear weapons, terrorism, solar flares, earthquakes, accidents, and complications with proper waste disposal are all determining factors in this theory. [3]

Naïve public awareness bolstered by political and economic propaganda has led to a new resurgence of a pronuclear mentality. Climate change, depletion of resources, and shortages of energy are all supposed justifications for such a mind-set. [4]

Since it is assumed that nuclear power itself, does not contribute to global warming, many misinformed individuals consider this a green technology. However, what nobody's telling them is that uranium mining is also a significant factor in the creation of greenhouse gasses. And regardless of global warming, true green energy technology should never carry with it the risk of Armageddon, let alone dangerous radioactive contamination! [5]

While analyzing the nuclear paradigm, five separate types of risk become immanent:

1. The contingent relationship of the nuclear fuel cycle and nuclear weapons.

2. Nuclear accidents: spills, leaks, contamination, meltdowns, and system failures.

3. The safe disposal of nuclear waste.

4. Terrorism: Dirty Bombs or an EMP Attack: An Electro Magnetic pulse attack could bring down the grid, thus, disabling backup generators and causing a melt down! Dirty bombs could be manufactured with low level nuclear waste and be detonated high in the atmosphere.

5. Natural Disasters: Earth Quakes, such as Fukushima that could trigger a meltdown, or Solar flares which are a serious threat to the integrity of the "Grid."

In the past, scientists working for industrialists, tycoons, and defense contractors, developed nuclear energy with the false hope that in the future a safe and permanent system of radioactive waste disposal would be developed. Even in the new millennium, scientists still don't fully understand the terrifying implications in the future of storing radioactive by-products. Nuclear waste has an afterlife which dramatically increases with age, peaking many thousands of years into the future, putting the fate of the human race in an extremely precarious position. [6]

Many conscientious scientists feel that by just taking into account the mitigating factors of our current nuclear storage technology, even without factoring in the threat of nuclear annihilation, there may only be cockroaches left inhabiting the earth someday if no safe and permanent system of storage is implemented. The public awareness of these factors are recondite and difficult to fathom. [7]

The conception of storing this material in deep mines, such as the Yucca Mountain Mines in Nevada, has been thrown around for over fifty years, but never properly implemented. The cost of this proposed site has risen exponentially over the years.

The problem is that the storage life of this material under our current technology is roughly forty years, and then must be dug up and re-contained. Since nuclear power is more than sixty years old, this has been an ongoing and never-ending process. Even though a nuclear plant can earn 2 million dollars a day, this ever-mounting cost of storage will have to be paid for with our tax dollars for eternity! [8]

> "There has never been a nuclear dump which hasn't leaked from low to high level. There is no technology which can safely contain this waste."
>
> — (Scott Portzline: Nuclear expert and chief editor of
> The Three Mile Island Alert newsletter) [9]

Scientists still lack the knowledge to properly estimate the potential migration of radioactive nucleoids into the environment. However, it has been determined that peak levels of radiation from this waste will occur thousands of years into the future, thus potentially displacing major portions of the earth's population, or even worse, making the planet completely uninhabitable! [10]

Economic pressures, combined with a more sophisticated mind-set about nuclear waste disposal (Who wants radiation in their backyard?), will be a motivation in the future to store these dangerous by-products in less developed countries. Many corrupt third world governments are notorious for cutting deals with rogue nations who would love to get their hands on this extremely hazardous material! [11]

A new world resurgence of nuclear power would make weapons of mass destruction much easier to acquire by these rogue nations. The spent fuel by-product of this dirty technology presents a Pandora's Box of its own! [12]

March 24, 2009: According to sources at CNN, the UK has recently reported an increased threat of chemical and nuclear terrorism. Officials in London believe that there are several other interrelated terrorist sources culminating and planning their next move. The Al-Qaeda Organization seems to be the predominant factor; however, there are many other factions of these Islamic extremists groups that, like Al-Qaida, share the same Extremist ideologies. [13]

"Surrogates of Hisbila, already boast about acquiring nuclear Weapons"
— (Salman Rushdie) [14]

Recent geopolitical events have further exacerbated this critical scenario. The United States, Iran, North Korea, China, India, Pakistan, Russia, and Israel all have major interests in the development of nuclear weapons. These political psychopaths hold the world hostage with their overzealous and megalomaniacal agendas! At this time, the Middle East and Northeast Asia represents a tumultuous hotbed of arms proliferation, each country trying to get an edge over the other in an attempt to gain superiority.

Nuclear weapons would be highly desirable by a weaker power, which either feels threatened or is currently under attack by conventional weaponry. [15]

The threat of acquiring dangerous radioactive by-products by nonnuclear nations led to international safeguards. Under the 1968, Nuclear Non-Proliferation Treaty, sovereign, signatory nations agreed to inspections by the International Atomic Energy Agency (IAEA). Currently the (IAEA) operates under the guise of the U.N. Security Council, supposedly adhering to international guidelines and policy regarding licensing, safety, compliance, and operational integrity of nuclear facilities all over the world. However, while the (IAEA) is supposed to be a Watch Dog for the nuclear industry, in reality, it is actually their Lap Dog! Their mission statement is to promote the nuclear industry! [16]

In 2009, Obama had won a peace prize for wanting to abolish Nuclear weapons globally. That was a nice gesture, but good luck implementing such a plan! [17]

In 2008 primary, the (infamously white supremacist and anti-Semitic), Lyndon LaRouche political organization, had opposed Obama for the election. It's not surprising that they were heavily financed by the Nuclear industry. Since the production of nuclear weapons is contingent upon the production of nuclear energy (by eliminating all nuclear weapons on earth, and thereby stopping any further production), then this would be a major blow to them. These Nuclear Nazis are truly purveyors of Hate and Armageddon! [18]

It was revealed in the release of a secret declassified document, that back in the early 60's, the air force came extremely close to mistakenly detonating an atomic bomb over N. Carolina which would have been 260 times more powerful than Hiroshima! Many years later, a B-52 flew over 1,400 miles from North Dakota to Louisiana, unknowingly carrying six nuclear-tipped cruise missiles! [19]

While looking retrospectively at these near disasters, the catastrophes suffered in Japan, Three Mile Island, and Chernobyl, along with the current draconian foreign policies we have today, then this could create an even greater worldwide propensity of terrorism and nuclear proliferation! The future of nuclear power appears sinister and foreboding! (20)

With the advancement of true green alternative energy, there should be no excuse to continue with this highly volatile and egregiously dangerous technology! For example, Ocean Power Technologies Corp. is a global provider of ocean-power-generating equipment with billions of dollars in projects all around the world. In PR Newswire (Oct. 13, 2009), OPT has teamed up with aerospace giant Lockheed Martin (a 29-billion-dollar corporation), in order to perform commercial engineering services in the development of OPT's tidal energy technology to be utilized in future large-scale applications.

> "OPT's proven Power Buoy technology uses 'smart' buoys, based on integrated patented hydrodynamics, electronics, energy conversion and computer control systems, to capture and convert energy from the natural rising and falling of waves into low- cost, clean electricity." [21]

Beyond OPT's patented technology which uses tidal power, there are many new methods being developed which can harness the energy of crashing waves. Experts have projected that these new systems could someday generate a potential target of two terawatts (2 trillion watts) [22]

Hydroelectric power is nothing new. Nicola Tesla designed the hydroelectric plant at Niagara Falls, which was built in 1907 and today safely generates an incredible 4.4 GWs of power. [23]

First Solar is a 12-billion-dollar corporation and the world's leading producer of photovoltaic technology. It has been established that solar power will grow by a staggering 26% annual growth rate to a projected 170–200 GWs by 2030. [24] & [25]

There are also untapped sources of wind energy. T-Boone Pickens called North Dakota the Dubai of wind generation. There are currently new buildings (domestically and internationally), already built, under construction, or in the planning phase, which incorporate wind turbine technology into their architectural designs. [26]

Sustainable author, Jeremy Rifkin, refers to the future of alternative energy technology is his book "The Hydrogen Economy." Hydrogen will someday be utilized as a viable fuel and may also replace nuclear technology someday with Hydrogen Fusion reactors which emit NO dangerous contaminants into the environment. [27]

Over the years, all these dangerous and uncertain factors of nuclear power generation had put a damper on industrial investors. This is the main reason why no nuclear power plants have been built in the United States for over twenty-five years. The infrastructure for such construction is either obsolete or nonexistent. Everything must be redesigned, retrofitted, or rebuilt. Interested utility investors are demanding heavy loan guarantees, expedient licensing agreements, and regulatory insurance. [28]

THE PRICE-ANDERSON ACT:

The creation of the Price-Anderson Act, last renewed in 1987, was designed to protect these pretentious industrialists. Under this Senate bill, the total liability incurred by the Nuclear Regulatory Commission (NRC), in the event of a nuclear disaster would be limited to only 8 billion dollars. Therefore, the taxpayers (you and me), would be forced to pay the difference. To put this into proper perspective, the Cher-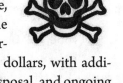nobyl disaster has already cost Russia over 300 billion dollars, with additional costs coming from interdicted land, radioactive disposal, and ongoing health effects which are mounting daily! The estimated cost of Fukushima could go into the trillions! [29], [30] & [31]

Scientists always want to believe that through gumption and fortitude, technological problems can be overcome by developing the proper applications and appropriate systems. However, the deeper that science delves into a nuclear solution, the more they realize the fruitlessness of such an endeavor. The conclusion always remains the same: too many unknown variables which can only lead to serious risks and grave consequences! What is logical and prudent for the survival and safety of mankind becomes obfuscated by greed and power. Neutron scientist Venkataraman Radhakrishnan writes:

> *"It is probably unrealistic to hope for a cooperative, compassionate political effort from predatory developed countries and their emulators which will address these options for the less favored."* [32]

FUKUSHIMA UPDATE:

The Fukushima Daiichi nuclear power disaster represents the worst radioactive catastrophe in history, rivaling that of Chernobyl by 400 times! Directly after the earthquake (50 minutes prior to the tsunami), reactors 1-3 automatically attempted to shut down. Contingent upon this reaction, emergency generators were activated to power cooling systems and electrical components. Subsequently, a 33 foot retaining wall, built to protect these backup generators from flooding, was breeched by a 46 foot tsunami! Previous cutbacks in funding had mitigated safety standards and strength of construction.

> *"The disaster disabled the reactor cooling systems, leading to releases of radiation and triggering a 30km evacuation zone surrounding the plant; the releases continue to this day!"*

> *"The failure occurred when the plant was hit by a tsunami that had been triggered by the magnitude 9.0 Hoku earthquake."*

Subsequent to this Melt Down (in 2011), 300-400 tons of dangerously contaminated radioactive water has been leaking into the Pacific Ocean every day! The entire Pacific Ocean has become contaminated, jeopardizing international fishing interests and threatening the health of populations from Hawaii to Alaska, the west coast, and all the way down to S. America! The radiation is not going away and will only get much worse over the next 100 years! [33] & [34]

At this time, NO viable plan has been initiated to stop this ongoing contamination! The Japanese have proposed building an ice wall, but this seems impractical, costly, and may not even be effective! The major problem is that they can't even get a robot into the melt down area to inspect! It's so radioactive at the point of the meltdown, that anything they send in, in order to assess the situation, gets fried! So how are people ever going to stop it?

"Fukushima was a harbinger of doom for all nuclear power plants!"

— (Arnie Gunderson)

The banks loaned money to pay the Japanese nuclear power plant work-ers, based on the premise that the plants would be reopened. This automat-ically creates a conflict of interest, where safety standards may be compro-mised in order to bring the plants back to full functionality, thus jeopardizing the health and safety of workers and the community!

Arnie Gunderson, Nuclear engineer and author of "Fukushima Cover Up," states that Japan currently has 100 times the naturally occurring rate of thyroid cancer! He goes on to say that 100,000 to one million people in Japan will contract cancer in the future from this disaster!

"Nukes are the Dinosaurs of the 21st century!"

— (Arnie Gunderson)

In an article recently published by author Neil M. Pine, in JetSetting-Magazine.com, about the dangers of nuclear power, he not only qualifies the effects from Fukushima, but also expounds about dangerous levels of contamination which came from the San Onofre nuclear plant, in southern California, which subsequently led to its permanent closure! (Also refer to link for the article about San Onofre, by Yoichi Shimatsu)

> *"These preliminary field studies at San Onofre and Catalina indicate that Edison power is primarily responsible for the kill-off and injuries to marine mammals along with possible illnesses among the human population of Southern California. This conclusion does not exonerate the nuclear op-erators at Hanford or Fukushima for their roles in the larger West Coast radiation crisis."*

— (Yoichi Shimatsu: Rense.com) [35]

HOW IS FUKUSHIMA DIFFERENT THAN CHERNOBYL?

Unlike Chernobyl, Fukushima was designed for the cooling systems to be contingent upon access to ocean water. There are other nuclear fa-cilities in the N. American which also utilize the ocean or large bodies of water, such as the now defunct, San Onofre California plant, or several other plants build on the Great Lakes.

Chernobyl was completely different than Fukushima. A series of explosions destroyed the reactor, releasing a cloud of radiation and contaminating a major portion of Europe. The reactor itself exploded while still active. At Fukushima, a magnitude 9 earthquake disabled the plant's cooling systems when a tsunami breeched the retaining wall which contains the emergency backup generators, leading to the melt down and subsequent on going contamination of the entire Pacific ocean!

Can We Have Another Disaster Like Fukushima In North America?

Unfortunately, in addition to a scenario like Fukushima, there are also several other major factors which also leave many U.S. nuclear plants vulnerable to such a catastrophe. Author Matt Stein, author of *400 Chernobyl's*, warns the public about the dangers of solar flares and EMPs (Electro Magnetic Pulse), attacks in regard to nuclear technology.

The manipulation of energy through a central power system (The grid), becomes highly exposed! The Grid represents major socioeconomic infrastructure affecting food services, distribution, internet and telecommunications, government and military, transportation, water, utilities and services, garbage removal, oil refining and gas pumping.

Solar Flares:

Over the past 152 years, the earth has been bombarded by roughly 100 solar storms (Also referred to as X-Flares), which have historically led to what are referred to as "Extreme GMDs" (Geo Magnetic Disturbances). These "Extreme GMDs" have only taken place twice during this 152 year time period. Today, a GMD of this magnitude could cause massive failures of 100s of nuclear facilities; a disaster far worse than Chernobyl and Fukushima combined! A Total Disaster that our civilization will not be able to afford!

A severe solar storm could take down the grid, causing widespread power outages and multiple nuclear power plant melt downs! There are measures which can be initiated in order to protect the grid from such a scenario, however, the way our bureaucratic system works, it may probably be too late by the time they incorporate any type of failsafe system. In the case of a GMD; after only one or two hours, each reactor backup generators will either fail to start or just run out of fuel!

EMP (ELECTRO MAGNETIC PULSE) ATTACKS:

EMPs can be utilized as a military weapon by launching a small nuclear device, like N. Korea already has, and what Iran is trying to get, which could be detonated above the earth's atmosphere, triggering what is referred to as *"The Compton Effect,"* splitting electrons off from atoms in the upper atmosphere, setting up a static discharge, that when striking the earth's surface, could blow out the entire nations grid in 1 second! All of our nation's power infrastructure could be instantly destroyed, causing massive melt downs at 100s of nuclear facilities! [36]

BOYCOTT, OR MAKE YOUR VOICE HEARD TO CORPORATIONS AND ORGANIZATIONS WHO SUPPORT NUCLEAR POWER:

Organizations:
- Lyndon LaRouche Organization
- American Nuclear Society
- Department of Energy
- Institute of Nuclear Power Operations
- International Atomic Energy Agency (IAEA)
- Nuclear Energy Institute <u>Corporations</u>:
- Bechtel
- S.C. Edison
- General Electric (GE)
- Dominion Resources
- Duke Energy
- NRG Energy
- South Carolina Electric and Gas (SCE&G)
- Southern Company
- Unistar Corporation
- Westinghouse Electric

Special Note:
Author Neil M. Pine, published an abridged version of this chapter in *Vision Magazine* in early 2010. Subsequent to this article, on Earth Day, April 24th 2010, he delivered a powerful speech on non-sustainable practices (warning people about the hazards of nuclear energy), at The University of California at Riverside (UCR), and received an Eco Hero Award. Only one year later, we had The Worst Man-Made Nuclear Disaster in History at Fukushima!

"The elimination of all antediluvian energy technologies (coal, oil and nuclear), and the development of truly sustainable energy sources (wind, solar and hydroelectric), will someday represent a monumental epoch in the evolution of humanity!"

— (Neil M Pine)

Notes:

[1]   Radhakrishnan: A world-renowned veteran neutron scientist. Trained at the Radium Institute in Paris, he questions the integrity of U.S. and foreign policy on nuclear proliferation and believes that the risk heavily outweighs the reward regarding the nuclear power industry.

[2]   The late John Gofman: Was one of the founding fathers of the antinuclear movement and was considered the most prominent critic of nuclear power in the United States.

He worked in the nuclear industry since 1939, helping to create the first atomic bomb (the Manhattan Project) used in WWII at Hiroshima. While working for the Atomic Energy Commission (AEC), he did extensive research into the effects of low-level radiation; and in 1969, he reported his honest but foreboding findings to the Joint Committee on Atomic Energy. Shortly thereafter, he claims that he was threatened at a special meeting by U.S. Representative Holifield, who was chairman of the joint committee which oversaw the AEC. Another congressman at that meeting was Melvin Price, one of the originators of the infamous Price Anderson Act! Congressmen Price, along with Holifield, intimidated Gofman with a threatening and inflammatory tirade, stating, "What the hell do you guys think you're doing interfering with the Atomic Energy Commission's program." According to Gofman, at the end of this conversation, Holifield leaned toward him and said, "Listen there have been a lot of guys before you who tried to interfere with the AEC program, we got them and we will get you!"

[3]   Comments on Nuclear Power by Stanley Thompson, Eugene, Oregon, 1997. (www. rathouse.org/radiation/CoNP/index.html).

[4]   Prescription for Survival: A Doctor's Journey to End Nuclear Madness (BK Currents (Hard

Cover)) by Bernard Lown, MD, and Howard Zinn, PhD, July 2008.

[5]		The Fate of the Earth, by Jonathon Schell. "It may be one of the most important works of recent years . . . there may still be hope to save our civilization" (Walter Chronkite, Avon Books, Hearst Corporation, New York, NY, 1992).

[6]		The Invisible Nuclear War: The effects of low-level radiation, The Massive Government Cover-Up, and the Continuing Battle Waged by the Nuclear Powers against All Life on Earth, by Carol Brouillet, Leslie J. Freeman, and Dave Ratcliff. Major excerpts courtesy of Dr. Rosalie Bartell, Dr. John Gofman, Dr. Jay Gould, Norman Solamon, and Dr. Earnest Sternglass (an unpublished text, photocopied by the primary author).

[7]		Environmental and Ethical Aspects of Long-Lived Radioactive Waste Disposal, Proceedings of an International Workshop Organized by the Nuclear Energy Agency in cooperation with the Environmental Directorate.

[8]		Uncertainty Underground: Yucca Mountain and the Nations High-Level Nuclear Waste by Allison M. Macfarlane and Rodney C. Ewing, April 28, 2006.

[9]		http://r.search.yahoo.com/_ylt=AwrSbmIL.iNWZ.cADgBXNyoA;_ylu=X3oDMTEydjlsMD-J0BG NvbG8DZ3ExBHBvcwMxBHZ0aWQDQjA5MzVfMQRzZWMDc3I/RV=2/RE=1445227212/RO=10/RU=http%3a%2f%2fwww.tmia.com%2f/RK=0/RS=wWG5AmDvdQwatxqwsPNiUVj9J4I

[10]		Nuclear Waste Stalemate: Political and Scientific Controversies, by Robert Vandenbosch and Susan E. Vandenbosch. University of Utah Press, Aug. 20, 2007.

[11]		Not in My Backyard: The Hand Book, by Jane Anne Morris. (Dedication: "To the Inhabitants of the Next Century . . . Who Will Surely Have Lots of Questions for the Survivors of This One.") Silvercat Publications, San Diego, CA, 1994.

[12]		Nuclear Power Plants as Weapons for the Enemy: An Unrecognized Military Peril, by Bennet Ramberg, University of California Press, Studies in International and Strategic Affairs, William Potter, editor, Center for International and Strategic Affairs, University of California, Los Angeles, CA, 1985.

[13]		Http:// news. bbc. co.uk_news/ 7960466. Stm.

[14]		Http:// www.hbo.com/billmayer/episode/2008_09_12_ep123.html.

[15]		The Gift of Time: The Case for Abolishing Nuclear Weapons, by Jonathan Schell. Metropolitan Books, Henry Holt and Company, New York, NY, 1998 (First Edition).

[16]		Info@IAEA.org.

[17]		"Obama Seeks to Abolish World Nukes," World News: 4/5/09- Jennifer Loven.

[18]		http:// en.wikipedia.org/wiki/Lyndon_ LaRouche http:// geocities.com/webcoast/LaRouche2.html.

[19]		B-52 Incident: www.democraticunderground.com/. . . /duboard.php?.92.all.

[20]		Cult of the Atom. The Secret Papers of the Atomic Energy Commission, by Daniel Ford. Simon and Schuster, New York, 1982.

[21]		h t t p : / / w w w . o c e a n p o w e r t e c n o l o g i e s . c o m / . PR News Wire, 10/13/09: Ocean Power Technologies and Locheed Martin Developing Utility-Scale Wave Power System.

[22]		http://plantgreen.dicovery.com/tech-transport/surfers-planet-mechanics-oceans. html?-campaign=daylife-article

[23]		http://www.en.wikipedia.org/wiki/niagrafalls.

[24]		http://www.firstsolar.com.

[25]		http://www.physorg.com/news/153148344.

[26]		http://www.pickensplan.com

[27]		www.foet.org/books/hydrogen-economy.html.

[28]		10/14/08, "The Nuclear Option," featured on CNBC: Discusses regulatory restrictions and costs as well as storage and safety. neinuclearnotes.blogspot.com/ 2008/.../nuclear-option-cnbc.html.

[29]   NRC, (Nuclear Regulatory Commission),http:www.nrc.gov/.

[30]   Price Anderson Alert-Nuclear Information and Resource Services: Reactor Watchdog Project: NIRS Info: http;//www.nirs.org/alerts/06-27-2001/2.

[31]   Ecocide in the USSR: Health and Nature Under Siege, by Murray Fishback and Alfred Friendly Jr., Foreword by Lester Brown (contains a number of sections on Chernobyl), Basic Books Division of Harper Collins, Publishers, New York, NY,1993.

[32]   Radhakrishnan, Siliconeer Magazine, Vol. IX. Issue # 5/ May 2008/ Nuclear Power: An Awful Option.

[33]   W.Mascheek, A Rineiski. M Flad.V Kriventsev Gabrielli. K. Morita: 24 hours at Fukushima: A blow-by=blow account of the worst nuclear accident since Chernobyl by Eliza Strickland Posted 31 Oct 2011.

[34]   PhillipLipscy, Kenji Kushida, and Trevor Incerti 2013. The Fukushima Disaster and Japans Nuclear Plant Vulnerability in Comparative Perspective.

[35]   http://blog.jetsettingmagazine.com/environment/article-5-the-insidious-nature-of-atomic- power-by-neil-m-pine

[36]   http://rense.com/general95/sanofre.html

[37]   http://rense.com/general95/sanofre.html

[38]   http://www.truth-out.org/news/item/7301-400-chernobyls-solar-flares-electromagnetic-pulses-   and-nuclear-armageddon https://www.washingtonpost.com/business/energy/why-japan-is-dumping-water-from- fukushima-into-the-sea/2021/08/26/4648e788-0649-11ec-b3c4-c462b1edcfc8_story.html

So if anyone ever tells you that nuclear power is okay, then please give them a magnanimous single digit salutation, while telling them to "Go Nuke themself and then Glow Home!"

The Conscious Planet

COPYRIGHT © Neil M. Pine, 2020

"Accuse not Nature,
she hath done her part;
now do thou but thine."

— (Milton, *Paradise Lost*)

CHAPTER 21

ENDANGERED SPECIES

(Illustration by Neil M. Pine)

INTRODUCTION:

Fauna and Flora

Due to man's ongoing foray with nature, we are facing one of the greatest extinction rates in history. Yet man, is the *"only one"* out of all these species who also possesses enough intellect to initiate a solution for this plight! The production of livestock still remains as the predominant factor in the destruction of our earth's ecosystems! For the rain forest and all its indigenous creatures who call this magnificent and incredibly beautiful, but fragile and rapidly depleting, environments home, there will be an ongoing struggle for survival into the twenty-first century! Endangered species journalist, Craig Kasnoff, in his 2019 exposé entitled, *Wildlife in a Warming World*, claims that up half of all animal and plant species, which inhabit some of the most critically endangered biomes on the planet, could become extinct by the turn of the century! [1]

This literature captures this dilemma of all endangered plants and animals, suggesting that if *man* does not transcend his anti-environmental, political, and economic protocol long enough to create a sustainably global solution, then he may be next on the endangered species list!

However, even in the midst of a maelstrom of overwhelming scientific evidence, the public still displays a vapid and pathetic mentality in regard to their awareness toward the environment and sustainability!

> *"The notion that science will save us is the Chimera that allows the present generation to consume all the resources it wants, as if no generations will follow."*
>
> — (Kenneth Brower, 2010) [2]

> *"The significance and scale of the global human footprint is not in doubt. Consumption of living resources as raw material and sinks for waste materials is high and growing!"* [3]

> *"The result of these transformations are almost all universally negative in their impacts on the biosphere."* [4] & [5]

> *"We are, he said, in the midst of one of the great extinction spasms of geological history."* [6]

In 2017, Kamran Nayeri, documented the work of Edward Wilson. In 1992, Wilson noted that human activities had increased "background" extinction rates by between one hundred and ten thousand times.

Nayeri published his report entitled *"How to Stop the Sixth Extinction: A Critical Assessment of E. O. Wilson's Half Earth"* [7]

Featured in chapters 23-29 is a more detailed description of many popular or unique forms of endangered species.

CHAPTER 22

RAIN FORESTS IN PERIL

(Illustration by Neil M. Pine)

The word pulchritudinous is a complicated sounding word meaning beauty, which has a synonymous relationship with our precious rain forests of the world, in that they too are also very beautiful, but also, so very delicate and complicated on their global scale of biospheric presence. This macrocosm of biodiversity represents plants, animals and indigenous people, who make their homes in the most complex, beautiful and endangered places on earth! The tropical rain forests play a key role in the basic functioning of our planet, as well as hosting 50% of all living species (plants and animals), in the world. They represent a priceless, pristine bastion of bountiful resources, only found in 3 parts of the world; Central and South America, Africa and south East Asia. Yet today, our precious rain forests are among the most threatened eco systems on the planet! Scientist have estimated that roughly 80 thousand acres of rain forests are destroyed every day! Along with this destruction, thousands of species face extinction.

Currently, there are officially 5,875 endangered species of plants alone. Cattle are still the predominant factor in rain forest destruction and desertification, resulting in a critical loss of flora! (Refer to chapter 7; "Dust Drought and Desertification")

While deliberating about the intricate biodiversity of rain forest environments, author Jeremy Rifkin writes:

> "So complex and so little understood are these tropical ecosystems that no more than two dozen scientists in the world are even competent enough to study their dynamics, according to Dr. Peter Raven, chairmen of the National Research Council's Committee on Research Priorities in Tropical Biology." [1], [2] & [3]

Never Take Trees for Granted: As we all know, the production of oxygen in the earth's atmosphere is contingent upon trees and other forms of flora, which in turn, also helps to eliminate carbon dioxide. One of the most outstanding examples of endangered plant life is right in the author own back yard in northern California. The behemoth redwood trees, which staggering robust growth, stood the test of time, through cataclysmic weather patterns and natural ecological disasters over thousands of years. Some of these trees have lived up to 4000 years, until man came along and literally *cut them down to size*, and what few are left are being threatened by air pollution!

Some Facts about Trees from the Arbor Council: A normal size tree generates enough oxygen to keep a family of 4 breathing for a year. Trees and plants also provide many other benefits besides oxygen. For example, only three trees strategically placed around buildings can cut the cost of air conditioning by up to 50%! Trees also increase the value of real estate. Statistics indicate that homes surrounded by trees can sell for up to 18-25% more than houses without trees. Trees are sustainable, biodegradable, and recyclable. They generate jobs and raw materials for thousands of uses in everyday life. We literally *could not live without them*!

Air Quality: Statistically, just by planting 20 million trees, we could generate an additional 260 million tons of oxygen and at the same time, eliminate 10 million tons of CO_2. The pollution from the average gas combustion vehicle can be absorbed in one year by growing four hundred trees. Trees provide better air quality. Leaves filter the air we breathe by removing dust and other particulates. Leaves absorb the carbon dioxide from the air, forming carbohydrates which are utilized in the plants structure and function. During this process, leaves also absorb other harmful pollutants such as ozone, carbon monoxide, and sulfur dioxide, while giving off oxygen. They provide protection, food, and shelter for a multitude of species.

Water Quality: Trees protect the integrity of water's purity from their leaves to their roots, while simultaneously trapping pollutants. Trees also help hold the soil in place along steep slopes. By slowing down this process of falling and running water, trees allow water to sink into the soil and help replenish underground drinking water supplies. Runoff of water fro forested areas was statistically 17% less than from developed areas. Not only do trees store water, but they also mitigate possible runoff which could lead to flooding. Trees growing on the banks of streams keep water cool for various fish and animal life.Trees are great natural mood elevator; they bring tranquility to our lives by reducing stress and lowering blood pressure. Employees also become more productive when they are exposed to trees while looking out the office window, or when driving to work. People recovering in hospitals who have views of trees have been proven to heal faster, use less pain medication, and actually leave the hospital earlier than those patients who only had a view of a brick wall.

People will spend more money at shopping malls with a lot of trees. People are actually willing to pay more for the same merchandise and shop for longer periods of time in shopping districts where trees are prevalent. [4]

For a complete list of all endangered plants, please refer to the following web address: www.earthsendangered.com/index.asp

Notes:

[1]   http://rainforests.mongabay.com/0812.htm.

[2]   http:// rainforests.mongabay.com/0801.htm.

[3]   Jeremy Rifkin. Beyond Beef, 92

[4]   http://www.dnr.wa.gov/arbor/facts/index.html.

TheConsciousPlanet.org

©Neil M Pine 2020

CHAPTER 23

ENDANGERED AMPHIBIANS

(Illustration by Neil M. Pine)

INTRODUCTION:

Amphibians are four-legged vertebrates which possess the versatility to survive on land or in water. Like fish, they are born with gills, but will develop lungs upon maturity. Another phenomenon associated with these creatures is that they can actually breathe through their skin. Due to modern industrialization (rapid urbanization) of emerging economies, such as India and China, along with climate change and environmental pollution over the years, almost one-third of all amphibian species have already become extirpated. With an additional 2,294 on the endangered species list, then this represents roughly another one-third of all species of amphibians still left in existence! In the past twenty-five years, 4 species have gone extinct and another 120 are facing extinction. Their survival is critically important to support a vast worldwide complex of ecosystems. While the extinction of amphibians in key local vicinities is of growing concern, human intervention will still remain as their greatest challenge into the future.

Some of the most critically endangered amphibians are as follows: the Chinese giant salamander; the Sagalla caecilian; olm; the Chile Darwin frog; the Betic midwife toad; the Gardiner's Seychelles frog.

CHINESE GIANT SALAMANDER

The Chinese giant salamander is the world's largest salamander, growing up to 1.8 meters in length; however, most specimens are considerably smaller. Their skin is rough, wrinkled, molted, and blotchy. Their color varies slightly from a muddy brown to a muddy green or black.

Geographical Habitats: The Chinese giant salamander is indigenous to mountain streams in China found at elevations below 1,500 meters along the tributaries of the Pearl, Yellow, and Yangzi Rivers. [1]

Dietary Traits: The giant salamander is a nocturnal creature, relying on sense of smell and touch to locate its prey. This prodigious amphibian acclimates itself to muddy, dark, wet rock crevices along riverbanks, feeding on fish, worms, crayfish, snails, and even other smaller salamanders. [2]

Endangered Status: The giant salamander is endangered primarily due to habitat loss, pollution, and human greed. The deforestation in China creates soil erosion, and the runoff causes the rivers to accumulate large silt deposits which provide unsuitable habitat for the salamanders. In certain areas of China, the building of dams has also had a significant effect on salamander populations, as it also changes or obstructs the natural flow of water, making it non-conducive toward salamander productivity. The Chinese giant salamander is also hunted for their meat, as it is, unfortunately, considered a Chinese delicacy. In addition, scientists believe that pesticides, fertilizers, and other industrial pollutants are impairing the salamanders' immune and reproductive systems, making them sick and further driving down their populations! [3]

The Chinese giant salamander is now protected from international trade by status of appendix 1 of the Convention on International Trade of Endangered Species (CITES).

Originating in the 1980s, fourteen nature reserves were established which protects more than 877,000 acres under the Chinese conservancy. [4]

THE SAGALLA CAECILIAN:

This wormlike salamander is only indigenous to an area about the half the size of Manhattan Island, living in an isolated section of Southeastern Kenya called Sagalla Hill. These ancient amphibians have a linage going back 370 million years to the Devonian Period.

Physical Traits: With its glib physical attributes and lack of limbs, the Sagalla caecilian has an appearance like a large earthworm. There is a pronounced pigmentation of the skin displaying an unusual pinkish-red tinge over a brown base. They live underground, burrowing through the dirt in a process which is referred to as undulation. In this way, they are also able to locate prey. Scientists have noted difficulty in studying these creatures by way of their subterranean lifestyles.

Dietary Traits: The Sagalla caecilian will feed on termites, earthworms, and a variety of soil-dwelling invertebrate. They may use different techniques to obtain their prey with a "sit and wait" or a "stealth" strategy, which allows them to sneak up on their prey and then seize it.

Endangered Status: It is difficult to determine the actual number of these creatures in existence today; however, their population trend is assumed to be in the decline and is rated "critically endangered" by the IUCN Red List of Threatened Species. Habitat destruction from farming interests seems to be their greatest threat, but chemicals and other farm pollutants are also a factor. Conservation International (CI) supports conservation projects in the region of Sagalla Hill through their Critical Ecosystems Partnership Fund (CEPF). [5]

PANAMANIAN FROG:

The Panamanian golden frog is an exquisitely beautiful gold-colored frog indigenous only to the mountainous regions of Panama. They primarily exist in streams adjacent to higher elevation rain forests. These creatures live on a wide selection of insects found near their streams.

Endangered Status: There are currently a variety of threats besieging these innocuous little creatures. As with most amphibians, they suffer from increased habitat loss, chemical pollutants, and illegal pet trade activities. However, in addition to these factors, scientists have recently discovered a deadly pathogenic fungus (Batrachochytrium dendrobatidis or chytrid fungus), which is also wreaking havoc with golden frog population and its habitats. It is officially listed as critically endangered on the IUCN 2007 Red List of Threatened Species. [6]

BETIC MIDWIFE TOAD:

The Betic midwife toad, also referred to as painted frogs, are indigenous to fragmented and isolated montane regions of Southeastern Spain, ranging in altitudes up to 7,100 feet above sea level. These creatures are quite unique in that after the female lays the eggs, the male will actually carry the eggs for over a month until fertile, hence the name midwife. Another interesting fact is that they are covered with tiny warts, which produce a highly toxic poison gas if the toads are handled or attacked. This natural defensive mechanism is so effective that the toads have almost no natural predators.

Dietary Traits: The Betic midwife toad's diet constitutes of various small insects and invertebrate, including crickets, worms, beetles, caterpillars, centipedes, and millipedes. However, the tadpoles subside on vegetable matter from a nutrient-rich aqueous environment.

Endangered Status: Populations of this species are extremely isolated, being more prevalent in the Alcaraz, Segura, and Cazorla Mountains, and are extremely rare in drier mountain regions which include Filabres, Baza, and Gador. Breeding habitats have been dramatically compromised due to drought conditions created by profligate water usage from cattle ranching and other agricultural concerns. Similar to the golden frog, the midwife toad is vulnerable to a fungal disease called chytridiomycosis. Due to these factors, in addition to the fact that their total habitat area only covers an area of two thousand square kilometers, they have been listed as vulnerable in the IUCN Red List of Threatened Species. [7]

THE OLM:

The olm is the only totally aquatic, cave-adapted invertebrate in Europe. This incredible little salamander can hunt for food in total darkness, relying on powerful senses of taste, smell, sound, and electro sensitivity. The olm is indigenous to Bosnia, Slovenia, Herzegovina, and Croatia. It was introduced to the French Pyrenees and northeast Italy and is believed to possibly exist in Serbia and Montenegro.

Physical Traits: The olm is elongated and slender, with three toes on the front limbs, and only two toes on the hind limbs. This creature has a whitish skin with a pink tinge. Another unusual physical characteristic of the olm is the absence of metamorphosis throughout its family tree, never maturing into adult form, but rather retaining their larval characteristics throughout life, complete with feathery gills, no eyelids, and a tail fin. They are highly adapted to subterranean darkness, giving them the sixth sense of electro sensitivity, which, like sharks, helps them to detect electric fields to catch prey. However, they don't appear to have the appetite of a shark.

The average length of the olm is only 230–250 mm, with the female slightly out sizing the male. They live to almost sixty years.

Dietary Traits: The olm will ingest detritus as well as various forms of cave invertebrate such as crabs, snails, and insects. These creatures are referred to as *"the masters of starvation"* while amazingly being able to go up to ten years without any food.

Endangered Status: In 2015, Significant declines in olm populations had been documented; however, it is very difficult to determine exact figures due to the difficulty in monitoring subterranean environments. The main ecological challenges facing the olm today are deforestation, which disrupts their sensitive subterranean ecosystems. Modification of adjacent agricultural land mass seems to also be contingent upon influencing olm habitat. Another factor that comes with tourism is water pollution, which is known to compromise their immune systems and reproductive organs. These creatures are very sensitive to toxins in water. They are also hunted as part of a black market pet trade. The olm is included in the Slovian IUCN Red List of Threatened Species. [8]

For a complete list of all endangered amphibians, please refer to the following web link: (www.earthsendangered.com/index.asp)

Notes:

[1]   http://www.arkive.org/chinese-giant-salamander/andrias-davidianus/#text=Range.

[2]   http://www.arkive.org/chinese-giant-salamander/andrias-davidianus/#text=Biology.

[3]   http://www.arkive.org/chinese-giant-salamander/andrias-davidianus/#text=Threats.

[4]   http://www.arkive.org/chinese-giant-salamander/andrias-davidianus/#text=   Conservation.

[5]   http://www.edgeofexistence.org/amphibians/species_info.php?id=548.

[6]   http://www.philadelphiazoo.org/zoo/Meet-Our-Animals/Amphibians/Frogs-and-   Toads/Panamanian-golden-frog.htm.

[7]   http://www.edgeofexistence.org/amphibians/species_info.php?id=601.

[8]   http://www.edgeofexistence.org/amphibians/species_info.php?id=563.

The Conscious Planet

CHAPTER 24

ENDANGERED BIRDS

INTRODUCTION:

In the avian category, there are currently 2,122 species of endangered birds. Every year, there are increasing numbers of endangered or threatened birds. While human populations continue to rise into the future, there will be a greater demand for natural resources and land development which will intensify pressure over their global ecosystems. Besides the jeopardization of their habitats, pollution also decreases their chances of survival. Out of all species of birds on the planet today, one in eight faces global extinction!

THE IVORY-BILLED WOODPECKER

The ivory-billed woodpecker is probably one of the most critically endangered species of bird on the planet. Once thought to be entirely extinct for many years, these birds have subsequently been spotted in extremely rare sightings, which have been confirmed in Arkansas and Florida.

Physical Description: The ivory-billed woodpecker is the largest species of woodpecker in North America. They carry a distinctive coloration made up of white patches over their shoulders and upper wings and also being adorned with a decorative- looking red patch on the side of their head.

Endangered Status: Other than a few anomalous sightings, the ivorybilled woodpecker is considered to be virtually extinct. The main factor against this species over the years has been habitat loss. Their biology is contingent upon the existence of very large and mature trees. The ivorybilled woodpecker is listed as critically endangered and possibly extinct by the International Union for the Conservation of Nature (IUCN). [1] & [2]

The California Condor

In 1897, the California condor was considered extinct in the wild. The last six birds known to be remaining in captivity at that time were placed in a "Captive Breeding Recovery Program."

Today, the California condor is still considered one of the world's rarest bird species, existing only in areas of reintroduction, making up range lands, oak savanna, rocky open scrub land, and mature montane regions of Baja California and Arizona. In 2005, the Habitat Conservation and Planning Branch of the Department of Fish and Game estimated that population levels had achieved up to 270 birds, including 145 already in captivity. Since that time, due to ongoing conservation efforts and public awareness, condor populations had risen to 463 as of 2017.

Endangered Status: The California condor is currently listed as critically endangered by the International Union for the Conservation of Nature (IUCN). [3]

Asian Crested Ibis

Except for one small region in China, these magnificent, extremely rare, and endangered birds are currently extinct throughout all ranges of their previous indigenous habitats. The Asian crested ibis sports a red masklike facial skin, which also matches their legs. They have a gray head, neck, mantle, and scapulars. Fully mature, they stand about twenty-two inches tall.

Dietary Traits: The Asian crested ibis feeds on crabs, frogs, small fish, river snails, other mollusks and beetles. Their main feeding habitat consists of rice fields, riverbanks and reservoirs.

Geography: Historically indigenous to far Eastern Russia, Japan, and China, this species has become extinct in all previous ranges except for the Shaanxi Province in Central Mainland China.

Endangered Status: Under CITES appendix 1, this species is legally protected under international law. It has also been officially listed as endangered on the IUCN Endangered Red List. In 1981, only seven of these birds were known to exist in the wild. With aggressive conservation efforts, as of 2017, more than five hundred had been identified. [4] & [5]

KAKAPO

The kakapo, also called the owl parrot, is a beautifully green colored flightless bird endemic to New Zealand. This species is unique in that it is the world's only flightless parrot and is also one of the longest-living birds, reaching up to 120 years in age. Besides their brilliant plumage, the kakapo displays a large gray beak, short legs, wide feet, with wings and a tail of relatively short length.

Dietary Traits: According to a research study conducted in 1984, twenty five plant species were identified as food for the kakapo. Some of the items in their diet include plants, seeds, nuts, fruits, pollen, and tree sap.

Endangered Status: Conservation protocol for the kakapo by the New Zealand Wildlife Service was established in 1950. Since then, ongoing efforts toward species reintroduction have been instrumental. By the early '70s, despite the earlier New Zealand Wildlife Service's agenda, the kakapos extinction status was still questionable. Thereafter, scientists initiated expeditions to several islands in New Zealand where they subsequently confirmed an estimated existence of more than one hundred as of 1977. In 1989, the Kakapo Recovery Plan was initiated, thus creating the Kakapo Recovery Group in order to implement this conservation effort. Since the plan's inception, the populations have been steadily rising. By 2016, their population had grown to over150, and as of 2019, due to optimal conditions that year, the Kakapo populations has reached over 200! The IUCN has officially classified the Kakapo as critically endangered on their Red List of Endangered Species. [6] & [7]

AMERICAN BALD EAGLE

The American bald eagle is probably one of the most iconic living creatures of all time. They are symbols of heritage, freedom, and majestic beauty, affording a history steeped with myths and legends. Since the beginning of humanity, the eagle has always been symbolic in our anthropological development. Their image depicted on currencies, calendars, representing many governments and countless organizations. During the '70s, due to DDT creating interference with their reproductive process by weakening the eagles' eggshells, this species almost became extinct! However, due to Environmental Protection Agency (EPA) regulations, education, and conscientious public awareness, this species is no longer threatened. Even though currently not endangered, due to its threatened past and strong public support for its survival, the American bald eagle still deserves honorable mention.

Habitat and Geography: The American bald eagle is mainly indigenous to the North American continent, covering most of Canada and Alaska, Northern Mexico, and the entire contiguous United States. They can be found adjacent to large bodies of open water and mature montane regions, which provide sufficient habitat for hunting prey and for nesting. [8]

For a complete list of all endangered birds and updated information, please refer to the following web address: www.earthsendangered.com/index.asp

Notes:

[1]   http://www.birds.cornell.edu/ivory/aboutibwo/.

[2]   https://www.audubon.org/news/possible-ivory-billed-woodpecker-footage-breathes-life-extinction-debate

[3]   www.answers.com/topic/california-condor.

[4]   https://www.edgeofexistence.org/species/asian-crested-ibis/

[5]   http://www.birdlife.org/datazone/speciesfactsheet.php?id=3801.

[6]   http://en.wikipedia.org/wiki/Kakapo.

[7]   lhttps://en.m.wikipedia.org/wiki/Kakapo

[8]   http://www.baldeagleinfo.com/.

The Conscious Planet

CHAPTER 25

ENDANGERED FISH

INTRODUCTION:

Armed with an onslaught of high-tech equipment—like sonar, infrared, and nets big enough to swallow twelve 747s—commercial fishing fleets can scoop up 80–90% of a given fish population in one year. What is even a more disturbing statistic is that for every one pound of shrimp caught, up to twenty pounds of various sea life (including birds, marine mammals, turtles, and other fish of limited commercial value), are killed and thrown back into ocean! Also, one-third of the entire world's fish catch is fed to livestock! Millions of years of checks and balances through predator prey relationship are virtually destroyed overnight! Commercial fishing remains as the predominant factor in global depletion of fish populations and should be recognized as a highly non sustainable practice.

Not only is commercial fishing environmentally non sustainable, but statistically speaking, it's not even economically sustainable. Even with all this plundering of our oceans, e.g., in 1994, commercial fishers spent 124 billion dollars to catch fish that were only sold for 70 billion! The remaining deficit of 54 billion dollars was paid for with government handouts coming from our tax dollars!

In a research study conducted by biologist Dr. Boris Worm, he warns of a worldwide collapse of the oceans ecosystems. Virtually every corner of the earth's oceans has been impacted by commercial fishing. This loss of bio-diversity is contingent upon the plundering of the world's oceans by com-mercial fishing interests. In 2014, Dr. Worm was regarded as a highly cited researcher by Thompson-Reuters. [1] & [2]

Could our precious oceans become what Earth Save Organization refers to as the "Next Dust Bowl?" But on top of all this, global warming, directly related to cattle, also plays a significant role in the production of oceanic methane plumbs, as well as drought, which leads to the destruction of nat-ural freshwater fish habitats, and whatever little clean water is left becomes polluted by fertilizer runoff or petrochemicals.

The chemistry of the ocean is also being dramatically influenced by ex-cess fertilizer runoff which adds nitrogen compounds to sea water. This influx of nitrogen compounds from fertilizers are responsible for massive plankton blooms, which deplete oxygen content from vast areas of ocean, thereby killing many fish, crustaceans, and shellfish. [3]

Other factors threatening fish populations are oil spills and petrochem-ical toxins, like dioxins and PCBs. In addition, heavy metal toxins such as mercury and cadmium, by-products of industrial production and other chemical toxins are also major threats to fish populations.

According to the IUCN, nearly 5% of all species of fish are on the critical-ly endangered Red List, facing extinction!

Not only does eating fish patronize non sustainable practices, but contrary to popular belief, fish can be just as high in cholesterol as chicken or beef. Also, the Physicians Committee on Responsible Medicine warns people about en-vironmental toxins in fish such as heavy metals, PCBs, and dioxin, along with dangerous pathogens, which can also affect human health. [4]

The statistics published by the Center for Disease Control indicates an average rate of 325,000 food poisonings annually, all directly linked to con-taminated sea food. [5]

Another disturbing fact is that government inspectors only inspect 1% of domestic catch and 3% of the imported catch for chemical and bacterial contamination. [6]

In 2017, a report published by the U.S. Environmental Protection Agency (EPA), nearly half of all fish tested were contaminated by excessive levels of bacteria from human and animal feces. [7] & [8]

Over the last sixty years, fish populations have been decimated by 90%! As of 2019, we have been deluged by various forms of oceanic pollution, involving fertilizer runoff (creating algae blooms), oil spills, radiation, heavy meatal toxins and fecal contamination. As if these aforementioned problems were not already bad enough, commercial fishing can also wipe out fish populations. Not eating meat dramatically reduces water incidence pollution. No longer indulging in fish, helps to preserve the natural balance of the ocean, thus protecting many species. Eating fish, is not just unhealthy, but its abstinence may also play a key role in saving the planet, while also exercising a little compassion, which is something that is lacking in our modern culture today. [9][1]

Here is an introduction to some unique endangered fish.

Rainbow Parrot Fish

The rainbow parrot fish is one of the most exotic and beautifully colored fish in the world found from Florida to Bermuda and from the Bahamas to Argentina. The male species can reach lengths of almost four feet. Their brilliant greens, blues, with orange fins give them their rainbow appearance.

Diet and Habitat: The rainbow parrot fish makes its home in the crevices of coral reefs. They are socially orientated, traveling in schools of about forty fish. Parrot fish graze on coral much in the same way that cattle graze in a

---

1   If someone does catch one of these endangered fish, please try to follow some simple procedures. First, release them without harm if possible; and if you wish to be more conscientious, then make a note of when, where, how many, what size they were, and send this information to wildlife authorities along with any photos or film.

field. Due to their highly calcareous diet, just one parrot fish can turn one ton of coral into sand within a year.

Endangered Status: These creatures are becoming increasingly more vulnerable due to loss of habitat which critically affects their mating and life cycles. These factors include overfishing, pollution, coastal development, and loss of food source. They are listed as vulnerable by the IUCN (1996). [10]

THE SPEARTOOTH SHARK

The Speartooth shark, also known as the Bizant River shark or the Queensland River shark, is a moderate-sized shark, primarily located in waters off the Western Pacific Ocean.

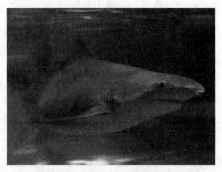

Physical Traits: It has a wide round snout with small eyes, a uniformly

Habitat: They are an extremely rare freshwater shark. Their survival being contingent upon large tropical tidal rivers and estuaries to which they make their homes.

Dietary Traits: They eat most fish, primarily prawns, gobies, jewfish, and bream.

Endangered Status: The main threat of the Speartooth shark is fishing and habitat degradation. They are protected by the Environmental Protection and Biodiversity Conservation Act of 1999 and are classified as endangered by the IUCN. [11]

PALLID STURGEON

The pallid sturgeon, also called the white sturgeon, is indigenous to many regions throughout the North American hemisphere. They have a very prehistoric appearance, earning them the reputation of being the "ugliest fish in North America." They can be generally found in the Missouri and Mississippi rivers. They are a generous-sized fish, reaching lengths of five feet and weighing up to eighty-five pounds.

Habitat: The pallid sturgeon prefers the environments of sandy or rocky bottoms of large turbid, free-flowing rivers. They are typically bottom feeders, scanning the river floors for insects, fish, and aquatic vertebrates.

Endangered Status: The pallid sturgeon has been officially listed as endangered in Arkansas, Illinois, Iowa, Kansas, Kentucky, Louisiana, Mississippi, Missouri, Montana, Nebraska, North Dakota, South Dakota, and Tennessee. They were classified by the IUCN as endangered in 1990. [12]

BLUEFIN TUNA

Highly desired commodities of sport and commercial fishing. Reaching lengths of ten feet and weighing up to 1,400 pounds, these hearty tuna can fetch up to $100,000 each! Due to overfishing, pollution, and climate change, these fish have become critically endangered.

Scientists are warning that without "prompt intervention," this species faces certain extirpation! One of the only reasons which have saved them thus far is that they migrate over thousands of miles, giving them a better chance as moving targets. However, with modern fishing technology, they stand little chance of escaping man's clutches! Even with international government sanctions, if there is a sushi bar that will buy it, then there is a fisherman who will catch it!

Geographical Habitat: They can be found in the Western Pacific Ocean, Eastern Atlantic Ocean, Mediterranean Sea, the Black Sea and the Caspian Sea.

Dietary Traits: The Bluefin tuna live off small fish and invertebrate consisting of sardines, herring, mackerel, squid, and crustaceans. [13]

THE BOCACCIO ROCKFISH

Out of the seventy species of rockfish living off the west coast of the United States, the Bocaccio rockfish, also known as a salmon grouper or a grouper, is one of the most endangered. Due to climate change and changes in ocean currents, these fish have become highly threatened. Even with conservation measures currently in place, scientists believe that it could take up to one hundred years for the Bocaccio populations to recover. The IUCN currently has a critically endangered rating on this species.

Geographical Habitat: The Bocaccio rockfish can be found from Alaska to Central Baja California, being most prolific from Northern Baja California to Oregon. They have a wide range of ocean depth in which they exist.

Physical Traits: The rockfish grows up to three feet long and may live as much as forty-five years. The Bocaccio rockfish have been monitored at various depths from surface levels to over 1,550 feet below sea level. Females grow faster and live longer than the males.

Dietary Traits: They primarily eat small fish and invertebrates such as shrimp, crab, anchovies, sardines, other small rockfishes, and squid. [14]

For a complete list of all endangered fish, and the latest updated information please refer to the following link: www.earthsendangered.com/index.asp

So if you believe that there is something fishy going on in our oceans, then abstain from eating all seafood and become a part of …

Notes:

[1]   http://www.naturalnews.com/025818_overfishing_Boris_Worm.html.   [2]/   https://en.m.wikipedia.org/wiki/Boris_Worm

[3]   https://www.npr.org/templates/story/story.php?storyId=128823662

[4]   The Physicians Committee on Responsible Medicine, http://www.targetofopportunity.com/pcrm.htm.

[5]   Centers for Disease Control and Prevention: http://www.cdc.gov.

[6]   https://wwwnc.cdc.gov/travel/yellowbook/2020/preparing-international-   travelers/food-poisoning-from-marine-toxins

[7]   The Consumers Union: http://www.peeplo.com/Consumer+Union/.

[8]   https://www.epa.gov/report-environment/consumable-fish-and-shellfish#condition

[9]   https://www.weforum.org/agenda/2018/07/fish-stocks-are-used-up-fisheries-   subsi-dies-must-stop/

[10]   http://earthsendangered.com/index_asp.

[11]   http://earthsendangered.com/index_asp.

[12]   http://earthsendangered.com/index_asp.

[13]   http://animal.discovery.com/adventure-fishing/sustainability/top-10-   mostendan-gered-fish-11.html.

[14]   http://en.wikipedia.org/wiki/Bocaccio_rockfish.

CHAPTER 26

ENDANGERED SPIDERS AND INSECTS

INTRODUCTION

Even though spiders and insects are not in most regards cosmetically appealing, lovable, or necessarily welcomed by mankind, they still provide a critical role in nature's system of checks and balances regarding agricultural pest control, soil amelioration, pollination of crops, and providing food for birds, reptiles, and other small animals. Insects are the backbone of earth's biological infrastructure; without them, we could not exist! So it's not easy to gain public interest and support to save some slug in New Guinea. You usually never see posters of endangered insects adorning the wall of people's homes, yet the survival of certain species of insects are just as critically important as the most prolific of endangered animals.

People need to differentiate their goals and vision when it comes to planetary custodial care. Are they involved with helping endangered species just because they are cute, or are they primarily interested in saving the planet? These are questions we all need to ask ourselves as we all work together toward a sustainable future. If people are sincere in their concern for the environment and wish to practice conservation, then they may need to reevaluate their goals as to why they want to protect endangered species.

There are approximately 725 species of endangered insects and thirty species of endangered spiders. However, scientists feel these figures are vastly understated. While insects make up 72% of global animal diversity, the International Union for Conservation of Nature (IUCN) and the United States Fish and Wildlife Service (USFWS) only have 7%, and 4%, respectively, of all the endangered species they list as insects. Even though people refer to them as such, spiders are technically not insects. In the phylum of arthropoda, there are several classifications, two of which being insects, and aradnida, spiders. This means that insects and spiders are related, but not the same. Some of the main distinctions are that insects only have six legs as opposed to the spider's eight. Another major difference is their physical structure. Insects have three body sections—the head, abdomen, and thorax—while the spider has two, the cephalothorax and abdomen. Another main difference is that most insects have wings, which spiders never have. [1], [2] & [3]

BUTTERFLIES AND MOTHS

Due to their sensitivity to environmental degradation, butterflies and moths are considered by science to be an *"indicator species,"* which generally indicates their well-being contingent upon the quality of their habitat. Their inherent sensitivity stems from their dramatic metamorphic life cycles. While transforming from caterpillar to a butterfly, these insects become exposed to a wide range of environments: from a herbivorous crawler to a winged nectar-gathering insect.

Endangered Status: Butterflies and moths have officially been listed as an endangered threatened insect species in the United States (2006) in the Threatened and Endangered Species System (TESS) published by the Department of the Interior, U.S. Wildlife Service, 2006. [4] & [5]

HONEY BEES

The Most Important Living Beings on the Planet!

In 2019, in the latest meeting at the Earth Watch Institute with the Royal Geographical Society of London, it was officially declared that Bees are by far the "most important living beings on the planet!" Without the honeybee, the entire base of our agricultural infrastructure would be destroyed! It has been estimated that more than 150 billion dollars' worth of crops would be lost if there were no bees to pollinate. Obviously, this would not help our diets, or the cost of food, which would skyrocket! In 2006, beekeepers started reporting major declines in bee populations. Initially, healthy bees were abandoning their hives in droves. Albert Einstein said "If bees disappear, humans would only have 4 years to live!" Research scientists have termed this mass exodus phenomenon as CCD (colony collapse disorder). In the latest EPA report, as of 2019, it had been determined that 90% of wild bee populations and almost 41% of all managed honeybee colonies, have already vanished! [6]

A new breakthrough study was recently released by the European Food Safety Authority (EFSA), which documented the hazardous affects that the pesticide Clothianiclin is having over bee populations. According to their report, this chemical was labeled as an "unacceptable" danger to bees. Monsanto's "Roundup" and other neonicotinoid based herbicides have also been linked to this phenomenon. [7]

It is of paramount importance that science immediately identifies the source of this species devastation. The U.S. Department of Agriculture has already allocated 20 million dollars toward research for conservation efforts in dealing with CCD. However, 20 million dollars is nothing compared to 150 billion! Thus far, the USDA has failed to find any viable solution. This author believes that banning the use of fungicides, herbicides, and pesticides by practicing organic farming, along with the abatement of burning fossil fuels, through sustainable agriculture production, would help to mitigate this problem. [8]

KAUA'I CAVE WOLF SPIDER

The Kaua'i cave wolf spider, also referred to as the blind wolf spider, is an extremely rare and unique spider, endemic only to three caves in the Koloa Poipu region of Kaua'i on the Hawaiian Islands. They are dark brown in color with yellowish stripes and reach lengths of a little more than three quarters of an inch. They are completely innocuous to humans.

Endangered Status: The Kaua'i cave wolf spider is officially listed as endangered under the International Union for Conservation of Nature (IUCN).

One major factor against them is that their entire habitat encompasses only a four square mile area, making them more vulnerable to extinction. Another problem they face comes from pesticide residues and other pollutants such as cigarette smoke. They are also threatened by natural predators like wasps and birds and also people. [9]

THE AMERICAN BURYING BEETLE

Description: The American burying beetle, also called the sexton beetle, is a large beetle with distinctive colorful markings on its back. Their bodies are shiny black with red-and-orange spots on their wings and frontal section of their thorax. They also can produce an unpleasant odor for defense purposes. They are called the burying beetle because they bury the carcasses of carrion in which they scavenge from surrounding habitat.

Habitat: Once widespread throughout the United States and Canada, this species today can only be found in isolated parts of Rhode Island, Arkansas, Oklahoma, Nebraska, and South Dakota.

Endangered Status: The IUCN has classified the American burying beetle as critically endangered. These beetles are also officially listed as endangered under the U.S. Endangered Species Act of 1969. They are becoming increasingly more threatened due to such factors as increased pesticide use, habitat fragmentation, and land development. They may also suffer from increased competition from other vertebrate scavengers, along with human collecting. [10]

For a complete list of all endangered insects, and the latest updated information, please refer to the following web address: www.earthsendangered.com/index.asp

Notes:

[1]   http://wiki.answers.com/Q/Are_spiders_insects.

[2]   http://www.highbeam.com/doc/1G1-179158846.html.

[3]   http://www.xerces.org/wp-content/uploads/2008/09/encyclopedia-article.pdf.

[4]   www.butterflys.org.

[5]   https://www.saveourmonarchs.org/blog/10-endangered-butterflies

[6]   New study shows @EPAgov should suspend use of the #pesticide that's killing bees.

[7]   http://www.gmoevidence.com/dr-huber-glyphosate-and-bee-colony-collapse-disorder-ccd/

[8]   https://www.sciencetimes.com/amp/articles/23245/20190709/bees-are-the-most-   important-living-being-on-earth.htm

[9]   https://en.m.wikipedia.org/wiki/Kaua%CA%BBi_cave_wolf_spider

[10]   www.amnh.org/nationalcenter/Endangered/beetle/beetle.html.

The Conscious Planet

CHAPTER 27

ENDANGERED MAMMALS

INTRODUCTION

Currently, there are 1,568 species of endangered marine and land mammals. Some of the more prolific or well-known species from this group includes manatees, killer whales, Hawaiian monk seals, blue whales, gray wolves, elephants margays, cheetahs, leopards, tigers, polar bears, giant panda bears, gorillas, and chimpanzees.

CHEETAH

They exhibit dazzling displays of speed; up to 64 MPH, making them the world's fastest land mammal. But they still can't outrun man's encroaching habitat destruction, mainly due to the production of livestock in Africa. This loss of habitat not only motivates persecution by ranchers but also creates competition between rival carnivores such as hyenas, lions, and leopards.

Over the past one hundred years, these creatures have suffered significant losses of population, leaving them vulnerable to extinction! It has been estimated, as of 2019, that less than 6700 cheetahs exist in the world today. They are listed on the World Conservation Union's (IUCN) Red List of Endangered Animals. They are hunted for their coats while also being killed for impeding livestock production. Without international sanctions, they would stand no chance of survival. Also, due to overhunting in certain areas, scientists have found genetic defects in cheetahs from interbreeding, which can increase cub mortality rates, create infertility, and impair their immune systems.

(Illustration by Neil M. Pine)

Physical Description: Cheetahs are easily identified by their black-spotted patterns with yellowish golden coats. There are some stripes which occur primarily at the tip of their tail and at the side of their eyes to the corners of their mouth. They reach lengths of up to four and half feet, excluding their thirty-inch tail. Mature cheetahs can vary in weights between 75 to 145 pounds, males usually being ten pounds more than females. Cheetahs usually can live eight to ten years in the wild, but have been reported to live up to seventeen years while in captivity.

Geographical Habitats: Cheetahs exist in small isolated populations indigenous to Sub- Saharan Africa and are very rare in parts of Somalia, Algeria, Niger, and have even been spotted in parts of Iran. They prefer grassy open range land and are not found in dense forested areas or wetlands. [1] & [2]

Polar Bear

With real estate, people know what it feels like to have your house underwater. But with global warming, these poor creatures' homes are literally underwater. Well, if your wife goes bipolar just because you missed a couple of house payments, then can you imagine running into a hungry nine-hundred-pound "bi-polar bear" who just lost its home? "Not a happy place!"

These animals' habitats are contingent upon ice formations to survive. Scientists believe that it is unlikely that the polar bears will be able to adapt to hunting land mammals. One group of scientists tracked a female polar bear and her cub. What they discovered was sad, shocking, and of grave concern! The bear and her cub started out by swimming 370 nautical miles through the Beaufort Sea in nine days. Thereafter, she intermittently swam and walked on sea ice for another 970 miles! After finally returning to the coast, near Kaktovik, where her original journey started, not only did she lose her cub but also 22% of her body weight!

These findings were documented by Alaskan scientists in December of 2008. Alaskan polar bear biologist Steve Amstrup, who has studied the Alaskan polar bears for several decades, stated that these conditions can only get worse as long as global warming continues to proliferate as scientists have warned us it would.

As of 2019, the polar bear is officially listed as a federal endangered species. The U.S. Fish and Wildlife Service has been trying to implement a system to "recover" polar bear populations as is mandated under the Endangered Species Act. However, with global warming being the contingent factor against their survival, scientists still haven't figured out how this can be accomplished! Even when this federal endangered classification was established, under the Bush administration, the law stipulated that it shouldn't be abused to make global warming policies!

The polar bear's main sources of food are ringed and bearded seals found in offshore waters off the continental shelf. But as the ice recedes in the summer months, it becomes increasingly difficult for the bears to hunt their prey. [3]

ELEPHANT (ASIAN AND AFRICAN)

Asian continents: the elephant, up until the early 20th century, had been heavily exploited for their tusks in the ivory trade. These mighty, yet highly vulnerable animals, have been victimized by man for hundreds of years! The legacy of their plight has been contingent upon poaching and loss of biodiversity.

These are the largest land mammals in the world. An African elephant can weigh as much as 8 tons! These magnificent creatures require vast amounts of land to roam and can consume hundreds of pounds of foliage in a single day. It becomes obvious that their survival in the future directly conflicts with the onslaught of modern civilization!

Statistically speaking, at the turn of the 20th century, there were approximately 3-4 million African elephants and about 100,000 Asian elephants. As of 2019, there were roughly only 450,000 - 700,000 African elephants and between 35,000 - 40,000 wild Asian elephants!

Elephants are very intelligent, as they can adroitly utilize their trunks for many purposes. They use it to bathe, drink, pick up objects, and as a loud trumpet to sound warnings to predators or communicate with the herd. [4]

Manatee

[5]

Habitats: Manatees inhabit coastal regions in shallow marshy areas by rivers and coastal inlets. They acclimate better to warmer climates, and unlike fish, they cannot survive below temperatures of 15°C (288°K, 60°F), relying on warm spring-fed water as a natural form of warmth during the winter.

Predators: Manatees have marginal incidents with predatory aggression. Sharks, killer whales, crocodiles, or alligators do occasionally prey on these aqueous cows, however, they do not pose any significant threat to their population.

Endangered Status: According to Florida Fish and Wildlife Conservation Commission, in recent years, there has been a dramatic decline in the manatee populations due to human-related issues. They cite some of the causes as habitat destruction, human objects obstructing or colliding, and natural causes due to temperature or disease. In 1996, nearly 20% of the entire world's population of manatees died due to a combination of red tide, boating accidents, and colder climates. However, many scientists believe that the manatees are trying to replenish themselves. Even so, there was a record 417 deaths in 2009, which almost rivaled the worst year on record in 2002. Propellers from boats seem to be their worst threat. Most manatees display scars from propeller marks on their backs. Marine mammal conservationists feel that the current environment presented to the manatees is inhumane, as up to fifty propeller scars have been reported from just one creature alone! Manatees have trouble distinguishing the low frequencies of larger marine vessels. However, in one study, when high frequencies were demonstrated, the manatees rapidly swam from danger.

Marine mammal veterinarians have organized and are initiating federal and statewide laws which would mandate humane standards.

Due to these ongoing conservation efforts, after 40 years of being classified as endangered, the Fish and Wildlife Service, in 2016, recommended downgrading the manatee's status from Endangered to Threatened. [6]

> *"The overwhelming documentation of gruesome wounding of manatees leaves no room for denial. Minimization of this injury is explicit in the Recovery Plan, several state statutes, and federal laws, and impact in our society's ethical and moral standards."* [7]

Killer Whale

A powerful and intelligent species of whale. They are a species of toothed whale, which can be found all over the world. Unlike the primarily warm-blooded sea mammals, like the manatee, killer whales are very diverse and can survive in any climate, from frigid Arctic oceans to tropical seas.

Their diets can also vary dramatically. While some may only feed exclusively on fish, others may prey on other sea mammals such as sea lions, seals, and walruses. And it's not unusual that they will kill common whales much larger then themselves. They have even been known to attack and kill great white sharks. This signifies these awesome creatures with a classification of "apex predators," which means they have no natural predators. Of course, man is their only threat due to prey depletion, habitat loss, and pollution mainly from PCBs. Animal psychologists have described killer whale populations as highly social, composed of "matrilineal" family groups, which are the most stable of any species. [8]

With advanced communication skills, along with sophisticated hunting techniques, killer whales display traits which animal behavioral experts have described as "manifestations of culture." [9]

BISON

Referred to as buffalo, were a prolific species of bovine, once reaching populations of up to 70 million on the North American Continent. They were victims of such a scourge that history and humanity will never forget! These hearty ungulates were indigenous to the North American Continent until their almost virtual extinction by Euro-American colonists in the late 1800s. As of 1889, there were fewer than one thousand left.

The bison were killed just for their hides, while others hunted the bison just for fun, killing them and leaving them to rot! The buffalo were sacred to the Indians for thousands of years and part of their ancestral heritage! This "Bovinical" genocide was a staggering blow to all Native Americans from which they have never recovered!

The buffalo provided for all their utilitarian purposes (food, shelter, and clothing). Many tribes during this era were credulous to the fact that all their precious bison had been eradicated. Some believed that they were hiding in caves just beyond the horizon. Rituals were conducted by the Indians to bring back their buffalo. However, tribal shaman priests needed buffalo meat as a sacrifice to complete this ritual. One such ceremony was referred to as "the anointing of the sacred pole."

Since there was no more buffalo, the tribe took whatever little money they had left, which had been collected from federal land transfers, and bought the white man's cattle as a substitute for buffalo in order to perform this ritual. Obviously, as history tells us, the cattle were sacrificed but nothing happened! They kept trying over and over again until all their money was gone.

Defeated, both spiritually and physically, they finally succumbed to sustenance farming in order to survive! [10]

In 1905, sixteen people formed the "American Bison Society" to protect and rebuild the captive population, and soon after, the government established the National Bison Range in Montana. There are an estimated 350,000 bison in North America as of today, 97% of them privately owned. [11]

Physical Characteristics: Unlike cattle, both the male and female buffalo have horns. And if this is not intimidating enough, they also have been described by naturalists as extremely dangerous and aggressive animals as compared to cattle which are typically complacent creatures. Males reach weights of up to two thousand pounds, sporting heavy horns, with a tremendous head and thick skull. The female cows typically weigh 40% to 50% less than the males. [12]

AFRICAN WILD DOG

The African wild dog, also referred to as the hunting dog, is an increasingly threatened species in East Africa. As a predatory species, the African wild dog is a contingent factor in elimination of sick and weak animals, thereby helping to maintain nature's checks and balances, while ultimately improving prey species. These animals have traditionally been given a violent and bloodthirsty stereotype, but as science understands more about the complexities of our ecosystems, this helps to mitigate this harsh image.

Physical Traits: Upon maturation, the African wild dog weighs between fifty-five to seventy pounds, reaching a height of approximately thirty inches at the shoulder. This animal has long legs, massive jaws, and pronounced bat-like ears. While these species resembles certain types of domestic dogs, they dramatically differ in that they only have four toes on each claw instead of five. They live between ten and twelve years in the wild. They have bushy tails with white tips, which they use to signal each other while hunting. Litters usually run about ten pups, but up to nineteen have been recorded.

Geographical Habitat: African wild dogs are indigenous to arid zones of the savanna, but can also be found in woodland and montane habitats of Africa where prey exists. They are more versatile than the cheetah which does not exist in wooded areas.

Behavioral Traits: African wild dogs exist in groups or packs of between six and twenty dogs. When the members of their pack fall below six, the effectiveness of their hunting becomes dramatically impaired. Before each hunt, the animals practice a bonding ceremony, where they will interact, vocalize, and touch each other, initiating motivation for the hunt. During the hunt, once prey is targeted, some dogs will follow close behind their victim while others will intentionally fall back, taking over the lead when the front runners get tired. Using this system, they can run at speeds averaging about 35 MPH.

As a pack, wild dogs are extremely efficient hunters; their targeted prey rarely ever escapes. According to behavioral scientists, wild dogs have generated high levels of intelligence and compatibility, by developing unique social skills involving body language and intricate forms of vocal communication.

Dietary Habits: Usually hunting in the early morning or late evening, wild dogs target gazelles or other antelopes, wildebeest calves, warthogs, rats, and birds. [13]

HAWAIIAN MONK SEAL

Under the category of "pinnipeds," as are other marine mammals such as manatees, walruses, sea lions, and other normal seals. Their Hawaiian name, "Ilio-holo-i-ka-uaua," means "dog that runs in rough water." Its common name (monk) is believed to originate from the shape of the head and body, giving it a medieval friar-like appearance. The name may also have some connection to the animals' solitary style of existence as compared to other species of pinnipeds that collect in large colonies.

Physical Traits: Adult male monk seals can weigh three hundred to four hundred pounds (140 to 180 kg) and reach lengths of up to seven feet. There is a genetic propensity for the females to obtain greater size than their male counterparts. Females can get to be eight feet long, weighing four hundred to six hundred pounds (180 to 270 kg). At birth, pups average thirty to forty pounds (14 to 18 kg). Their life expectancy ranges from twenty-five to thirty years. They have large black eyes which gives them greater underwater vision and a sleek body to cut through underwater turbulence.

Geographical Habitat: The Hawaiian monk seal is endemic to the Hawaiian Islands, meaning they can't be found anywhere else, unlike the word indigenous, which means that a species can also exist in other places. This endemic theory becomes prevalent due to the unique tropical, geographical habitats (climate and biological factors), of the Hawaiian Islands.

Behavioral Traits: The Hawaiian monk seal spends a vast majority of its life in the ocean, but does come ashore to give birth, molt (shed its skin), and find shelter in serious storms. Monk seals are extremely sensitive to human disturbance, also tending to live a much more docile life than the rest of their pinniped counterparts, who congregate in large groups.

Females take raising their pups seriously, nurturing them on shore for five to six weeks, where the female can lose hundreds of pounds, which is a very challenging process. So much so that most females cannot reproduce every year.

Dietary Habits: The monk seals search for food along the many coral reefs and sandy ocean bottoms which adorn the surrounding Hawaiian Islands.

They hunt for fish and invertebrates such as eels, flat fish, octopus, and lobsters. Monk seals seeking food have been known to dive to incredible depths of from three hundred to six hundred feet, which can last for up to twenty minutes.

Endangered Status: The Hawaiian monk seal is the most endangered marine mammal in the United States. It is on the critically endangered list, heading toward extinction! It is considered one of the most endangered mammals in the world. [14]

Aye-Aye

The aye-aye is a rodent-like-looking primate with dramatically large eyes and sharp teeth. It has been described as having "blazing eyes," spoon shaped ears which can rotate like "radar dishes" and "thin black hands with elongated fingers." This unique endemic creature can only be found in the rain forests of Madagascar. Up until the nineteenth century, aye-ayes were considered rodents. Like rodents, they have a dental structure with incisor teeth which grow exponentially. Another distinguishing feature is its tail, which is bushy and longer than its body.

Behavioral Traits: The aye-aye spends most of its time in or around trees. They sleep in the trees, making their nests out of branches and leaves. These animals, being nocturnal, are highly active during the night, covering long distances without rest. Another unusual fact is that at only five and a half to six and a half pounds, they are considered the largest nocturnal primate on earth. Most nocturnal animals are usually very small. Also they are the only primate which uses echolocation, similar to bats, to hunt its prey. They are gentle and completely harmless to humans, while one scientist noted that the only reason to put them in cages was so that they would not steal their writing utensils or paper.

Dietary Habits: Scientist believe that the aye-aye travel in pairs, while hunting for food. Behavioral experts believe this displays an intelligent and pragmatic approach to sustenance gathering. They forage for coconut, fruits, insects, and their larvae. The aye-aye uses its long middle finger to aid in this process by tapping it on the bark of trees in order to determine the movements of larvae or wood-boring insects or grubs.

Endangered Status: These creatures have been listed as critically endangered by the EN-US FSW on June 4, 1973. There is a terribly ignorant superstition of the people of Madagascar which portends foreboding implications for the future of this gentle primate. It is commonly believed that this innocuous creature actually possesses nefarious magical powers which can bring death and destruction to any village where it is seen. And unfortunately, these unique animals show no fear of humans and are usually killed on sight! Today there may only be as little as one thousand to two thousand of these creatures left on earth! Efforts have been made to propagate their populations, but education to the human populations of Madagascar is also of paramount importance! As of 2012, the official IUCN status was highly endangered. A population decline of approx. 50% or more, had been estimated over the last 35 years with and addition 50%, projected over the next 30 years. [15]

TASMANIAN FORESTER KANGAROO

Forester kangaroo is an endemic species of kangaroo only found in Tasmania. They are a subspecies of the eastern gray kangaroo also found only in Tasmania. Adults can weigh up to 130 pounds, reaching more than 6.5 feet in height, making them the largest marsupials in Tasmania. They are incredible jumpers, covering up to thirty feet in a single bound. However, when not jumping, they tend to use all four legs, their front limbs being much smaller that their hind legs, which can utilize five fingers for grasping objects. The Tasmanian Forester kangaroo exhibits a superior sense of smell, sight, and, also like the aye-aye, are able to swivel their ears for advanced hearing ability. They prefer to live in grasslands or open wooded areas.

Dietary Habits: The Tasmanian forester kangaroo are an herbivorous species, subsiding on a plentiful variety of grasses, herbs, leaves, and shrubs.

Behavioral Traits: As a social animal, they prefer to live in small groups referred to as "mobs," which typically include one dominant male, two or three females, and several more young males along with "joeys" (baby kangaroos). Male Forester kangaroos will display aggression in order to establish dominance over the "mob." This behavioral dominance incorporates a form of boxing competition. The mating rituals of the forester kangaroos are contingent upon the winner of this competition and therefore establish leadership. "Joeys," or baby kangaroos, can remain in the mother's pouch for up to three hundred days and will still nurse for another eighteen months after leaving the pouch.

Endangered Status: Since the 1800s, Tasmanian Forester kangaroo populations have declined by 90%, due to hunting and agricultural expansion. Their future survival faces such challenges as climate change, disease, poisoning, and livestock proliferation. The Australian government has established provisions to protect the kangaroo by setting up land "reserves" where the populations can grow undisturbed. [16]

MARGAY

Determined that the margay is more adopted toward arboreal (inhabiting or frequenting trees) living than any other cat species. They are the only known species of cat that possess the faculty to rotate its hind legs 180%, which enables it to run down trees headfirst like a squirrel. They are also known to be able to hang from branches by using only one hind foot. These animals have a life span of up to eighteen years.

Geographical Habitat: The margay is indigenous to evergreen forests throughout Mexico, Panama, Columbia, Peru, Paraguay, Uruguay, and isolated parts of Argentina.

Dietary Habits: Margays' primarily forage through trees for sustenance. Their prey consists of birds, eggs, small mammals, lizards, and tree frogs. Like normal cats, they will eat grass and various other vegetation to aid in digestion.

Endangered Status: The margay is extremely rare and endangered throughout their geographical habitats. Unlike domestic cats, the margay can only bear one kitten, which further exacerbates the propensity of their survival. Through the years, thousands of these beautiful creatures have been hunted for their coats, and even with international sanctions regarding their protection, illegal poaching still exists today. [17]

TIGER

Remained the subject of intrigue and mystery. Tigers have historically exhibited a profound impact on human cultural development, influencing the creation of myths and legends based on this creatures awe inspiring physical attributes. Depending on the context of the encounter, there still exists an element of ambiguity when judging people's reactions towards the tiger. They can be feared, distrusted, respected, or admired, especially depending on where you're standing! Folk legends display a dichotomy of beliefs between the tigers' power to save or destroy. In Sumatra, for example, the tiger is considered a friend who can be summand in times of sickness or calamity.

Physical Traits: Siberian tigers are the heaviest of all tigers, reaching weights of over 500 pounds. The lightest species of tiger being the Sumatran tiger, weighing in at about 250 pounds. Most tigers exhibit distinct physical traits: dark stripes, white stomachs, with a tawny, yellowish brown coat.

However, there are some anomalies which have been documented that represent unusual colors, including all black or all white. Tigers display six times better night vision than humans. They live up to ten years in the wild, but have been known to live twice as long in captivity.

Geographical Habitat: The living paradigm of the tiger is contingent upon three main factors:

1. Sufficient ground cover

2. Generous populations of wild prey
3. Indelible water supply

These habitat requirements can be accommodated through grass or swampland, but the tigers favor more heavily wooded areas. Tigers can be found from India to Southeast Asia and all the way to Siberia.

Dietary Habits: Tigers usually prey on large species of animals such as deer, wild buffalo, and wild pigs; but they have also been known to hunt fish, birds, and monkeys. On rare occasions, they may attack baby elephants, leopards, bears, or even other tigers!

Endangered Status: Tigers are officially listed as endangered by the IUCN. However, they are still illegally poached for their fur and body parts, as well as having to deal with habitat destruction! According to scientists, the Chinese tiger and the Siberian tiger are both facing threats of total extinction! [18]

GIANT PANDA

Mountainous regions of Central China. They are the Sichuan, Shaanxi, and Gansu provinces. At one time in history, giant pandas inhabited low-lying areas, but due to agriculture, infrastructure, and other land development, they remain restricted to only a few mountain ranges.

They live at elevations of five thousand to ten thousand feet in coniferous and broad-leafed forests which support a healthy undergrowth of bamboo.

Physical Traits: The giant panda is a black-and-white bear with a giant stuffed-toy-looking head. They have thick wooly fur which keeps them warm in cold climates. As an enigmatic design of nature, they display black patches of fur over their eyes and their ears. People perceive these bears as cute and adorable; however, don't let their looks fool you. They can be just as dangerous as any other bear! Giant pandas exhibit large molar teeth and strong jaw muscles for chewing tough bamboo. They are roughly the equivalent size of an American black bear, standing at two to three feet tall at the shoulder (when on all fours) and as much as six feet tall while standing upright. Males can reach weights of up to 250 pounds, which is about 30 pounds heavier than what females can achieve. Scientists have not determined how long a giant panda can live in the wild; however, it was documented at one Chinese zoo that one lived to be thirty- five. In addition, unlike bears indigenous to temperate climates, giant pandas do not hibernate.

Dietary Habits: The digestive track of a giant panda is actually similar to that of a carnivore, thus creating a process of insufficient assimilation, as food is passed before completely being digested. Therefore, to make up for this factor, the panda must consume 20 to 40 pounds of bamboo a day to achieve optimal nutrition. To reach this daily quota, the panda may spend from ten to sixteen hours eating and foraging. When not engaged in this practice, they can usually be found sleeping or relaxing.

Endangered Status: The IUCN (World Conservation Union) has official-ly placed the giant panda on the Red List of Threatened Species. There are only about 1,900 giant pandas left on earth; 1,600 in the wild, and 300 in captivity. The latest IUCN report in 2017 states that Panda Populations are rebounding, but they still are not completely out of danger. Many factors could negatively affect this outlook. [19]

CHIMPANZEE

Characteristics and behavior most similar to that of humans and is recog-nized as one of the most intelligent animals in the world. Scientific research has recently documented that these primates are the closest living relative that humans have. The DNA similarity between chimps and humans is 98.4%.*

In addition, they also use crude tools for digging, foraging, and grooming. Chimpanzees are believed to be more advanced than monkeys, belonging to ape family along with gorillas, bonobos, and orangutans. Their lifespan in the wild ranges up to fifty years, but chimps in captivity have been reported to live to sixty. [20]

Geographical Habitat: Chimpanzees spend as much time in trees as they do on land, mainly for eating and sleeping. They are indigenous to the tropical rain forests of Africa.

Dietary Habits: The chimpanzee is an omnivorous primate, their diet consisting primarily of fruits, leaves, buds and flowers of plants, berries, and seeds. However, not that long ago, people thought that the chimpanzee was a strict vegetarian, until forty years ago when animal behavioral expert
Jane Goodall observed wild chimpanzees hunting animals and eating meat at the Gombe National Park in Tanzania. It was documented that chimpanzees kill and eat up to 150 small- to medium-sized animals such as small monkeys, pigs, and antelope; however, this only comprises roughly 3% of their diets.

Endangered Status: The endangered status of the chimpanzee is contingent upon three main factors:

1.) Habitat Destruction: A little over fifty years ago, more than one million chimpanzees inhabited rain forests, encompassing at least twenty-five countries in Africa. As of 2008, as little as two hundred thousand were left in existence, with only six countries left still having the right environmental conditions in order to support these endangered primates. The other remaining nineteen countries have destroyed their natural rain forest habitats, replacing it with ranching and agriculture, building development, and highway infrastructure. By 2017, Chimpanzees, had already disappeared from 4 countries in Africa.

2.) Illegal Captivation: Chimpanzees are unlawfully seized in the jungles of Africa and sold as pets or to laboratories which may conduct unethical experiments or utilize their body parts for sale on the black market. Even though baby chimps seem like a great pet, these animals were never intended to be locked in cages their whole life. Animal behavioral experts have deemed this practice unethical, citing that after growing up in a bountiful and pristine rain forest environment and then being placed in a cage, this would represent deplorable and inhumane living conditions.

3.) Hunting and Killing Them for Meat: Many poachers, mostly natives of the African jungle, will kill these helpless chimps just to harvest "bush meat," or sell this meat to customers who are willing to pay exorbitant amounts of money for primate flesh. [21]

GORILLA

Due to their size, enormous strength, and erroneous media coverage, gorillas have traditionally been portrayed as dangerous, violent beasts. King Kong would be the best example. The truth is that gorillas are really very gentle and friendly toward humans. The only time they get angry and will feign violence is when their families are threatened. Like chimpanzees, the gorillas' DNA is closely linked to humans. They are omnivorous; however also like the chimp, their diets mainly consist of vegetation. [22]

Geographical Habitat: Gorillas primarily dwell in rain forest environments such as swamps, fields, and humid, lowland forest regions. While their populations are being increasingly threatened, they can still be found in regions of Nigeria, Cameroon, Gabon, Congo, Central Africa Republic, and Zaire.

Physical Traits: Gorillas have black or brownish grey fur, with black skin on their chest. Reddish fur on their head is quite common among Cameroon gorillas. As the male species matures, their backs turn a silvery color. When standing upright, a full-sized male adult can reach heights of up to 5'6" and 4'6" normal stance. The female gets to be 5" upright and up to 4" in normal stance. The male's weight ranges between 300 and 500 pounds, while the female goes from 150 to 250 pounds. The gorillas have an incredible arm span (up to 9'2" from one particular male subject). There exists three subspecies of gorilla living through different parts of Africa; however, their differences are insignificant. They are the western gorilla, the eastern gorilla, and the mountain gorilla. [23]

Endangered Status: It is roughly estimated that there are between ten thousand to thirty-five western gorillas living in the wild. There are only approximately four hundred wild eastern gorillas, and only as little as six hundred of the variety of mountain gorillas left. There are 550 gorillas in captivity worldwide. With the severely dwindling population of mountain variety gorilla, they are indeed facing certain extinction if something is not done to intervene with the current trend! [24]

For a complete list of all endangered mammals, and the latest updated information, please refer to the following web address: www.earthsendangered.com/index.asp

Notes

[1]   https://www.awf.org/wildlife-conservation/cheetah

[2]   http://nationalzoo.si.edu/Animals/AfricaSavanna/factcheetahs.

[3]   https://www.livescience.com/topics/polar-bears

[4]   https://www.google.com/search?client=safari&channel=iphone_bm&source=hp&ei=EPa-PX eXpFs7U-gTH5rQQ&q=pic+of+elephant&oq=pic+of+el&gs_l=mobile-gws-wiz- hp.1.0.0l8.3232.7451..9075...0.0..0.86.635.9......0....18..41j41i22i29i30j0i131j46i131j46j0i10.0vlWdMJC79U#imgrc=7mcns1X8LX2wVM

[5]   https://en.m.wikipedia.org/wiki/Manatee

[6]   https://www.fws.gov/news/ShowNews.cfm?_ID=35428

[7]   (Marine Mammal Medicine, 2001, Leslie Dierauf and Francis Gulland, CRC Press).

[8]   https://www.afsc.noaa.gov/nmml/education/cetaceans/killer.php

[9]   Rendell, Luke, and Hal Whitehead, (2001), "Culture in Whales and Dolphins," Behavioral and Brain Sciences 24(2):309-324. PMID11530544.

[10]   http://www.answerbag.com/q_view/3140.

[11]   "Beyond Beef," Jeremy Rifkin, 92.

[12]   http://www.fws.gov/bisonrange/nbr/facts.htm.

[13]   http://www.outtoafrica.nl/animals/engafricanwilddog.html?zenden=2&subsoort_id=4&bestemming_id=1.

[14]   http://www.mcbi.org/what/what_pdfs/sealFacts.pdf.

[15]   http://ielc.libguides.com/sdzg/factsheets/ayeaye/population

[16]   https://dpipwe.tas.gov.au/wildlife-management/fauna-of-   tasmania/mammals/possums-kangaroos-and-wombats/macropods/forester-kangaroo

[17]   https://www.wildcatfamily.com/leopardus-lineage/margay-leopardus-wiedii/

[18]   https://relay.nationalgeographic.com/proxy/distribution/public/amp/news/2017/12/palm-oil-sumatran-tigers-extinction-big-cats-animals

[19]   https://www.worldwildlife.org/species/giant-panda

[20]   http://www-bcf.usc.edu/~stanford/chimphunt.html.

[21]   https://wwf.panda.org/knowledge_hub/endangered_species/great_apes/chimpanzees/

[22]   http://www.buzzle.com/articles/endangered-gorillas.html.

[23]   http://www.koko.org/about/facts.html.

[24]   https://gorillafund.org/mountain-gorillas-tragedy-fragile-success/

The Conscious Planet

Chapter 28

Endangered Reptiles

Introduction

Reptiles appear to be quite prolific; we see them everywhere in our daily lives, from pet stores to the media. People just don't realize that many species of reptiles are facing the possibility of extinction! The three major obstacles facing reptile populations today are habitat loss, pollution, and human greed. Of the 502 species of endangered reptiles, the giant sea turtle and the Komodo dragon seem to be the most popular in terms of public awareness, but this does not necessarily make them any more significant (in terms of anthropogenic biology) than the other 500 endangered reptiles. In 2019, the Hawaiian, and the Pacific Island regions, had reported an overall improvement in sea turtle populations.

Green Sea Turtle

The green sea turtle, also referred to as the giant sea turtle, is an ancient species of sea reptile, its origins going back millions of years. These magnificent Cheloniidae are the largest of all hard-shelled sea turtles, with adults weighing up to 350 pounds, but documented history tells us that there was one recorded over 700 pounds. These turtles reach lengths of three to five feet and can live to be one hundred years old.

Dietary Habits: The green sea turtle is a mainly herbivorous creature, subsisting on sea grasses and algae, which is unique among other types of sea turtles. It is believed that what gives them their rich verdure color is their high-chlorophyll diets. However, besides plant matter, they may hunt for squids, crabs, and other small sea creatures.

Geographical Habitats: The green sea turtle can be found in tropical to temperate waters worldwide. They occupy sea grass beds and nesting land called mangroves during egg-laying periods, but they prefer the shelter and rich selection of foods found in coral reefs.

Endangered Status: Green turtles are threatened all over the world, but especially so in Florida and Mexico where they are facing extinction! It is difficult to estimate how many green turtles are left in the world; however, scientists have determined by studying migration patterns of female turtles that their populations have been steadily declining. The green sea turtles are protected by various international sanctions and treaties as well as many national laws. This species is prohibited under appendix 1 of the Convention on International Trade in Endangered Species of Wild Flora and Fauna (CITES), banning it from international commerce. Green sea turtles are also listed in appendices 1 and 2 of the Convention on Migratory Species (CMS), qualifying them under the protection of the Memorandum of Understanding on the Conservation and Management of Marine Turtles and Their Habitats of the Indian Ocean and Southeast Asia (IOSEA) and the Memorandum of Understanding Concerning Conservation Measures for Marine Turtles of the Atlantic Coast of Africa. In addition, there have also been provisions made for the green turtles protection under annex 2 of the Specially Protected Areas and Wildlife (SPAW). Under U.S. protocol, an international binding treaty dedicated solely to the protection of marine turtles was created: the Inter- American Convention for the Protection and Conservation of Sea Turtles (IAC). Also, green turtles are protected under the Memorandum of Understanding on ASEAN Sea Turtle Conservation and Protection, and the Memorandum of Agreement on the Turtle Islands Heritage Protected Area (TIHPA). [1] & [2]

Tuatara

The tuatara is an amazing prehistoric-looking reptile with a history dating back 220 million years. They are believed to be one of the oldest living species of living creatures on earth.

Dietary Habits: The tuatara is generally nocturnal, emerging from their burrows to eat anything that comes along, mostly insects such as worms, slugs, millipedes, and a huge species of bug, indigenous only to New Zealand, called Wetas. [3]

Physical Traits: The tuatara is said to be more primitive than any other reptile. This species has virtually remained the same for over 200 million years. The males can get up to twenty inches long and weigh over two pounds, while the females only reach weights of a little more than one pound. They sport a spiny crest on their backs made of triangular-shaped soft folds of skin, which is more pronounced in the males and can be stiffened and displayed for purposes of intimidation. The tuataras color scheme can fluctuate from an olive green to orange or reddish brown.

Endangered Status: The tuatara is so rare that they can only be found on several islands off the coast of New Zealand. On the Brother's Island, for example, it has been estimated that less than three hundred still exist.

They have been classified as an endangered species since 1895, and while their populations on the other islands may be more prolific, they are still threatened with extinction due to habitat loss and the introduction of non-indigenous predators such as rats, weasels and badgers. The tuatara is classified by the IUCN as vulnerable. [4]

Gharial

The gharial, also referred to as the long-nosed crocodile, is part of the family of gavialidae, a long-established group of crocodilians indigenous to India and Nepal.

Physical Traits: The gharial displays a protuberant or elongated snout which will narrow and lengthen with age. The bulging appendages referred to as bulbous growth on the tip of the male snout are called Ghara. This Ghara serves as a function in several ways; it produces a loud buzzing associated with social behavior, produces bubbles which aids in courtship, and is also a visual stimulus for attracting other crocodiles. The gharial is physically the second largest species of crocodilian, reaching lengths of up to twenty feet.

Geographical Habitats: The gharial primarily exhibits an aquatic lifestyle, preferring the calmer sections of deep fast-moving rivers. However, during dry seasons, they will accumulate along sandbanks formed by river deposits, which increases their risk of vulnerability. Another factor against them is their non-adroit perambulatory capability to negotiate land mass. When traversing across land, they cannot lift their bodies clear of the ground; this creates drag and dramatically impairs their movements. In water, however, these creatures are extremely agile, using their tail combined with webbed feet most effectively. [5]

Dietary Traits: Evolution has perfectly adopted the teeth jaws of this creature to predominantly eat fish. Their thin snout gives them superlative low-resistance underwater fishing capability. In rare circumstances, the gharials have been known to attack large land mammals, but usually don't hurt people. The juvenile is more adapted to eat insects or other smaller vertebrates such as frogs.

Endangered and Conservation Status: During the first half of the twentieth century there were up to ten thousand of these creatures inhabiting areas throughout India. By 1970, it was evident that dramatic population declines had taken place, thus prompting S. Biswas of the Zoological Society of India to initiate programs in nine protected areas where species propagation and conservation measures have been established. However, despite all these efforts, due to fishing, land development, and pollution, the gharial population has still been rapidly declining. In 2007, they were declared critically endangered on the IUCN Red List. In addition, the Gharial Conservation Alliance (GCA) was formed in 2007 by scientists, experts, and conservation stakeholders. [6]

Komodo Dragon

Komodo dragons, also referred to as land crocodile, are the heaviest lizards on the planet, weighing up to two hundred pounds and reaching lengths of ten feet. Usually females are about two feet shorter and fifty pounds less. They exhibit an excellent sense of smell which enables them to scavenge for carrion. Regardless of their size, these huge lizards are quite agile as they can climb trees and are good swimmers. Similar to a shark, the Komodo dragon's teeth are laterally compressed with serrated edges. Also like a poisonous snake, they  are venomous, which is also an important factor for disabling their prey. Their life expectancy is between twenty and forty years.

Geographical Habitats: The Komodo dragons are indigenous to the Sunda Islands of Indonesia, where they have thrived for millions of years, yet incredibly, man has only been aware of their existence for roughly only one hundred years. These volcanic islands provide these creatures with harsh, dry, and arid conditions, which is conducive toward reptile population. [7]

Dietary Traits: Like a dragon, the Komodo dragon has a voracious appetite, literally being able to eat up to 80% of its own body weight. They will eat anything moving, including acts of cannibalism. Typically, they prey on large pigs and small deer, but have even been known to kill full-sized water buffalo and small or injured humans.

Endangered Status: The dragon's greatest threat is volcanic activity which causes fire and drives away prey. Currently, tourism, along with the infringement of habitat by the poaching of prey species and also the killing of the dragon for its skin on the black market, are its greatest challenge.

It is estimated that between three thousand to five thousand Komodo dragons exist in the wild today. They are protected under the Convention on International Trade in Endangered Species (CITES). [8]

BOA CONSTRICTOR

The Boa constrictor is a large full-bodied, non-venomous snake indigenous to Central and South America and some islands of the Caribbean. Although the boa is not the largest snake, it can still reach lengths of up to fourteen feet in some regions of the world, achieving weights of over sixty pounds. Females tend to be roughly 25% larger than the males. The overall size of the boa is contingent upon subspecies, geographical location, and availability of suitable prey.

Dietary Traits: The boa constrictor's diet represents a wide variety of rodents, mammals, and birds. As the snake matures and increases in size, the size of its prey will also get larger. Larger animals such as ocelots have been reported to be consumed. The boa will often ambush its prey, lying in wait for the opportune moment to make its move. However, they will actively hunt in geographical regions where low prey concentration exists. These nonpoisonous snakes depend on their teeth for grabbing and their powerful body to crush its victim. Due to their slow metabolism, these snakes only need to eat as little as once a week or can even go up to a few months without food.

Endangered Status: Boa constrictors are protected under the Convention on International Trade in Endangered Species (CITES), appendices 1 and 2, depending on the subspecies. They are threatened mostly by human and other animal predation but are also endangered by exotic pet and snakeskin trade. [9]

For a complete list of all endangered reptiles, and the latest updated information, please refer to the following web address: www.earthsendangered.com/index.asp

Notes:

[1]   https://www.worldwildlife.org/species/green-turtle

[2]   https://www.sciencenews.org/article/endangered-green-sea-turtles-may-be-making-comeback-us-pacific

[3]   https://www.doc.govt.nz/news/media-releases/2019/tuatara-thriving-on-moutohora/

[4]   https://en.m.wikipedia.org/wiki/Tuatara

[5]   https://www.iucnredlist.org/species/8966/149227430

[6]   http://www.flmnh.ufl.edu/cnhc/csp_ggan.htm.

[7]   http://animals.nationalgeographic.com/animals/reptiles/komodo-dragon/

[8]   https://relay.nationalgeographic.com/proxy/distribution/public/amp/animals/reptiles/k/komodo-dragon

[9]   http://en.wikipedia.org/wiki/Boa_Constrictor.

The Conscious Planet

CHAPTER 29

WHAT IS; THE STATE OF THE ART LIFESTYLE?

PART 1

What is The State of the Art Lifestyle? If you asked a random group of people, they would come up with a myriad of answers. Some would say it represents opulent wealth: Jet Setting; breakfast in Tokyo, dinner in Paris, the biggest or most expensive homes, planes, boats or cars. While others would associate this statement with technology, using the smallest, lightest, fastest electronic gadgets with the most efficient memory and baud speed, to maximize performance and modernize their lives.

What The State-of-the-Art Lifestyle should really mean is the quintessential method of bio- consciousness. It represents civility and compassion for life, holistic concepts and respect for the environment (a true Green mindset). This symbiotic relationship with nature is good for our planet, our health, and our psychological well being. These three key elements are paramount toward our anthropogenic evolution. The State of the Art Lifestyle is the future of humanity! (Refer to chapter 2: The Psychology of the Cattle Culture)

According to Harvard scientists, in their Keynote Address, entitled Future of Life, it is scientifically explained how our carbon footprint is creating a cataclysmic impact on the earth's ecology. It is their contention that we would need four planets to sustain life if all seven billion inhabitants of earth were to emulate the egregiously profligate lifestyle of the average American! An ecological footprint represents a person's toll on nature. The dynamics of this paradigm are calculated by using 2 key factors. 1.) The amount of biological material consumed. 2.) The amount of carbon waste created. This quantitative calculation determines your footprint. In essence, we must ecologically overcome this foreboding dichotomy if we someday wish to achieve total sustainability.

However, The State of the Art Lifestyle is so much more than reducing this carbon footprint. It is a futuristic mindset of health and compassion for all living creatures. Not only can we dramatically reduce our footprint by becoming vegans, but we automatically become more empathetic and civilized human beings in the process. We must move across the archaic ethos of western dogma and with the vicissitudes of time, we will emerge as a truly compassionate and sustainable society of the new millennium!

"The philosophy of one century is the common sense of the next"
— (Henry Ward Beecher)

Therefore, in order for man to survive in the 21st century, he must change and adapt away from the antediluvian thinking of the past, and take on a more humble, practical, and ascetic quality of living. But the good news is that with technology and a little ingenuity, we can still have our *vegan cake and eat it too*; e.g. a 600 horsepower, 100% electric car. Mansions powered by solar, etc....

The legacy of our ostentatious and excessive lifestyles has traditionally painted a picture of the ugly American. But, with The State of the Art Lifestyle we can finally break free of this derogatory stereotype. We will gain respect and sovereignty from all nations of the world, especially when they begin to emulate our new green lifestyle. As Gandhi said "If we want to make a change in the world, we must become part of that change" The State of the Art Lifestyle gives strong affirmation to this statement. This new change also represents freedom: Freedom from the dependence on foreign oil, freedom to breathe fresh air, and freedom from guilt associated with the slaughter and exploitation of animals.

This freedom will translate into a new sense of pride and awareness in America, one that our forefathers never knew. Only 150 years ago, our forbearance justified the genocide of buffalo, the decimation and subjugation of indigenous Americans, and the enslavement of African people. We have evolved so much as a nation since this time; from Exxon to Edison, from Red Meats to Red Beets, from the Marlboro Man to the Zig Zag Man, from shots of Tequila to shots of Wheat Grass. This political correctness and State of the Art green, enlightened thinking will make us the new hero's and patriots of future generations!

Could we be entering into The Age of Aquarius, or are we merely teetering on the brink of Armageddon? In addressing this statement, there are as many varying viewpoints as the existence of human orifices. However, one thing is for certain; if we wish to preserve our way of life for future generations, we must do much more than change our light bulbs like those silly public service commercials would have you believe. Changing a light bulb merely scratches the surface of total sustainability.

Let's say we are fighting against each other on the battlefield, and the aliens arrive like in War of the Worlds. They proceed to destroy all earthlings, regardless of their political, religious or racial affiliation. Now, instead of expending natural resources to kill each other, all nations of the world would be forced to unite together toward a common goal to save the planet. Global warming represents those terrifying aliens, yet we are still blind to the cataclysmic implications of what this climate change holds in store for this planet! Our culture is so vapid and narcissistic, that we must literally see some ominous or threatening image, like monsters from outer space or a gigantic meteor headed at us, before people finally wake up and take notice that there is a serious danger threatening our very existence!

We must put our prejudices and antiquated traditions aside, and work toward a common goal to save the planet. Knowledge is the key. We must learn the truth, examine and understand the root of our problems.

There must be some form of public service announcement which exemplifies the destructive nature of livestock production, so that we can make a conscientious decision for our children and the future of humanity! We can't afford not to!

However, due to government (USDA), and corporate (Beef & Dairy Council) subreption (A malicious concealment of the truth), public service commercials never mention the drought, famine, soil degradation, water pollution, or global warming, created by the production of livestock. The truth must be told! The state of the environment is in jeopardy. *If people don't see it on TV, or hear it on the radio, then they just don't believe it*!

Prolific environmental author Jeremy Rifkin, writes in his 1992 publication, *Beyond Beef,* that cattle production is wreaking havoc on all 6 continents of the planet. He also states that people who live in agricultural regions around the U.S., are told to practice water conservation by their local municipal water districts. However, they are never informed that cattle are the main source of drought in their region!

> *"The devastating environmental, economic, and human toll of maintaining a world wide cattle complex is little discussed in public policy circles. Most people are largely unaware of the wide ranging effects cattle are having on the ecosystems of the planet and the fortunes of civilization. Yet cattle production and beef consumption now rank among the gravest threats to the future well-being of the earth and its human population."*
>
> — (Professor Jeremy Rifkin:
> The Foundation on Economic Trends)

This was written over 25 years ago, and it's more significant now than ever!

In addition, the Smithsonian Institute, Earth Save Organization, and The United Nations report Livestock's Long Shadow, just to name a few, all have credible scientific evidence as to the profligate and destructive nature of cattle production.

And after realizing these principles, movie star mogul, Leonardo DiCaprio, had backed the documentary film *"Cowspiracy"* in 2015, and is currently on a worldwide quest to eradicate livestock production.

Therefore, besides utilizing alternative energy and green transportation, *The State of the Art Lifestyle* also represents vegan ideals. Eating meat is not only terrible for the environment and your health, but it also patronizes evil and inhumane, Auschwitz style, slaughter house conditions.

"Does something have to die in order for you to live?" You have to be some kind of a *prima donna* to think so. There is a certain pride in knowing that you can live and thrive while not damaging the environment or killing other living creatures. Once you realize this inner truth an enormous burden of guilt may be lifted off your shoulders. Now you are truly living in harmony with nature and you may begin to feel an illumination of spirit. Living without this cycle of death we automatically become more humane, and therefore more human:

More Humane.................*Less Beastly*
More Compassionate.........*Less Barbaric*
More Gentle.....................*Less Brutal*

More Assertive.................*Less Aggressive*
More Ethical...................*Less Deceitful*

More Green!

PART 2
The Practical Application of The State of the Art Lifestyle
The State of the Art Health:

Diet:

1. Go Vegan: The enormous amounts of natural resources needed to produce meat and dairy products are one of the most significant causes of climate change. Raising cattle is environmentally irresponsible and represents the antithesis of true bio-sustainability. The food we eat, not only affects our health and the environment around us, but it also has a major impact on our psychology and the way we form our moral values. Make compassionate and bio-sustainable choices for yourself and your family. (Refer to chapter 2: The Psychology of the Cattle Culture)

The use of any animal products, even though supposedly acquired compassionately, such as milk, cheese, butter, and eggs, still patronizes the exploitation of animals, pollutes the earth, wastes natural resources and valuable land which could have been utilized sustainably. E.g., statistically speaking, on just one acre of land, you can yield 20,000 lbs. of potatoes. On that same land, within the same time frame, only 165 pounds of beef can be produced. Also, it takes 2500 gallons of water to produce just one pound of beef, but it only takes 6 gallons of water to grow one pound of broccoli. (Refer to chapter 9: 'Dust, Drought, And Desertification)

2. Bio-sustainable: A vegan diet consisting of locally and regionally grown organic produce represents an optimal awareness of bio-sustainability. Even if the produce is organic; if it had to be flown in halfway around the world, then it should not qualify as truly bio sustainable. However, it is still more environmentally conscious to eat organic produce grown out of the country then it is to eat locally raised meat. Of course, this is not to mention the compassionate aspect of this choice. (See chapter (Cattle and Egregious Greenhouse Gasses)

3. Macro-biotic:

305

a. No GMO's: Genetically Modified Organisms (herbicide residues). Genetic engineering creates a mutation between various species in an unnatural way. The USDA in collusion with the FDA, have irresponsibly allowed these laboratory designed mutations to enter our food chain without proper testing or mandatory labeling! This is an outrage! E.g., GMO corn, sugar beets and soy, currently on the market, contains high levels of Roundup herbicide residues and have been proven to cause non-Hodgkin's lymphoma, tumors, sterility, liver and kidney damage and a myriad of other conditions! On how to protect yourself The Implications of Genetically Modified Organisms. (See chapter 16 The Implications of Genetically Modified Organisms) [1]

b. No Artificial Sweeteners: Aspartame, Sweet n Low, Sucralose, and NutraSweet are pure poison! Only use natural zero, or low glycemic sweeteners such as Stevia, xylitol, agave, maple syrup, palm sugar, malt barley or *honey. (* Non Vegan) [2]

c. No Processed Foods: Avoid highly processed flours and sugars, usually found in cookies, cakes or breads. Besides genetically modified ingredients, containing Roundup herbicide residues and other dangerous herb and/or pesticides, also avoid other chemical additives such as artificial colors, artificial flavors, chemical preservatives, dough conditioners, hydrogenated oils, anti-caking agents and food stabilizers. Corporate America wants you to be addicted to these filthy additives! [3]

d. No Alcohol or Drugs (Illegal, prescription or over the counter): Avoid using any substance, natural or otherwise, that has toxicity which could potentially damage the kidney or liver. Almost all over the counter and prescription drugs, including aspirin, Tylenol and antibiotics, can seriously damage your internal organs, especially when being mixed with alcohol. [4]

e. No Irradiated Foods: This process is completely unacceptable, also having a severe impact on the kidney and liver by changing the molecular structure of food more to that of an inanimate object, therefore, dramatically increasing the shelf life, but making it much more difficult to digest. E.g., Plastic fruit looks beautiful, however, you wouldn't want to eat it! [5]

f. No Microwave Foods: Use a convection oven instead. Microwaving food takes all the life force out of it, turning your sustenance into empty calories, while also creating an acidic body PH which is conducive to cancer and the ravages of free radicals. Furthermore, the carcinogenic substance Benzene along with radiolytic compounds (which can't be found anywhere else in the universe), are also created during the microwaving process. [6]

g. Organic: Organic farming in conjunction with the proper crop rotation, fortifies the soil, leaving it fertile and chemical free. In fact, the Federal Gov. allocates tax credits for practicing land conservation to anyone who produces certified organic crops on their land.

Part of true bio-sustainability, is leaving the earth in the same condition as it was before it was used. In addition, organic produce can contain up to 4 times the vitamin and mineral content, and without the pesticide residue of its conventionally grown counterparts. Also, organic foods afford a much higher flavonoid count. There's nothing as flavorful as an organic peach picked ripe from the tree. Using non-organic chemical phosphate fertilizers, pesticides, fungicides, and herbicides, degrades and poisons the soil, while also robbing it of essential minerals and nutrients. (Refer to chapter 17: Organic Conventional vs. Conventional Foods)

h. Raw Juicing:

1. Wheat Grass: In the modern world, people are so caught up with taking the latest drugs. Billions of dollars every year are spent by the pharmaceutical companies, just on advertising alone. The average senior citizen over 65 years old is on 11 prescription medications. The irony is that the latest State of the Art drug is literally as old as dirt. Grass is as old as dirt. Mother Nature, in the form of wheatgrass is more beneficial than any drug ever created by man! Yet if you ask your doctor about it, they would probably say "What's that?" Wheat grass contains anti-neoplastic properties which rejuvenate cell tissues and have documented case histories of dissolving tumors. No man made drug, can or ever will contain anti-neoplastic properties. It is the life force behind this fresh juice which creates a natural panacea of healing. [7]

2. Misc. Fruit and Vegetable Juices: (see diet sec). Fresh squeezed juices are essential toward a healthy lifestyle.

Never drink soda, coffee, or other heavily caffeinated or carbonated beverages, as these products rob your body of calcium and may contain GM ingredients. Organic raw juices are loaded with vitamins, minerals, and anti- oxidants, which gives your body sustained energy while also helping to speed up your metabolism. Throw away your coffee maker and buy yourself a juicer!

Vitamin, Mineral, and Herbal Supplements: Over the years, due to modern commercial farming practices, our soil has become severely depleted and no longer affords the nutritional benefits it had once supplied. Therefore, even though it's better to get your vitamins and minerals from fresh food sources, it may still be important to supplement your diet. [8]

4. The State of the Art healing: (Chemical and drug free holistic concepts) a. Detoxification: Part of The State of the Art Lifestyle is living free from external and internal toxins. People need to change their diets and chemically oriented lifestyles, and return back to a more basic, wholesome, and health-conscious mindset! Take nothing toxic to the kidney or liver in order to cleanse and detox. This would include: All prescription, OTC, and illegal man-made drugs. Even aspirin or Tylenol can seriously damage your kidney and liver, especially if mixed with alcohol.

1. Colon Cleanse: Relieves putrefaction and cleans out inner intestine. (Colonics)

2. Candida Cleanse: If you have ever taken anti-biotics, then you are probably suffering from some form of this yeast overgrowth in the bowels. The problem with anti- biotics, is that it kills friendly, as well as unfriendly forms of bacteria. (see chapter on Argyrol and other colloidal silver products)

3. Parasite Cleanse: Opportunistic organisms can wreak havoc with your health.

4. Organic Chelating Therapy (Heavy Metal Cleanse): Helps to prevent Alzheimer's and other physical side effects due to toxicity from pesticides and exposure to volatile chemicals.

5. Petro-chemical Cleanse: Molecules from these byproducts culminate in your cell tissue. Avoid drinking water from plastic, especially if it's been heated up or it has been sitting in the sun! Also, stay away from cosmetics which may contain petroleum jelly or methyl parabens, as petroleum byproducts can be absorbed through the skin and into the blood stream. [9]

b.) Holistic Practitioners:

6. Holistic Doctors: & Herbal Medicine

7. Nutritional Therapists

8. Chiropractors: Chi-ropractors can help to enhance your Chi, (life force)

9. Massage Therapists
    a. Shiatsu
    b. Rolfing
    c. Acupressure
    d. Reiki [10]

    c.) The State-of-the-Art Compassionate Medicine:
      420/ Compassionate Health Care:
      For well over 100 years, the lumber, cotton, petroleum, pharmaceutical, alcohol, and tobacco industries have successfully lobbied to oppress, repress, subjugate, slander and propagandize the hemp industry. Hippies weren't the first American pot farmers to suffer undue persecution.
      Hemp was used for sailing ships, covered wagons, clothing, shelter, food, medicine, fuel, and even the first American flag.
      When newspaper mogul, Randolph Hearst (Who owned a major portion of the Northern California forest during the 1930's), found out that it costs four times as much to make newspaper from trees rather than from hemp, he initiated a public smear campaign, funding such propaganda films as "Refer Madness," and then financially backing the 1937 Marijuana Tax Act which crippled American hemp farmers and thus, classified marijuana as a narcotic like heroin!

Since 1937, more than half of the world's forests have been cut down to make paper!

"We will not permit American Farmers to grow hemp!"
        — (Gen. Barry McCaffrey, U.S. Drug Czar, 1999)

Today, for the first time in modern recorded history, tobacco has finally taken a back seat to cannabis in terms of political correctness. Our society has finally evolved from The Marlboro Man to The Zig Zag Man ! How many more gold medals could Michael Phelps possibly have won if he didn't smoke pot? Corporate America would not be so quick to judge if Michael Phelps was just drinking a beer! But because marijuana does not represent any particular corporation, he was being unfairly black-balled.

So, people expect that every time Michael Phelps is sore from working out, or has strained or sprained his muscles, that he should take substances like Tylenol, that are proven to cause bleeding ulcers, and/or kidney and liver damage in hundreds of thousands of Americans annually! Marijuana is probably the safest and most non- toxic form of analgesic.

The latest research from John Hopkins University, not only proves marijuana to be harmless, but also indicates legitimate medicinal properties for the treatment of (Cancer, Epilepsy, Pain, Glaucoma, Asthma, Insomnia, Ulcers, and Depression), that help people to heal naturally, without the harsh side effects of Chemotherapy or drugs, which cause kidney and liver damage. Even aspirin, Tylenol and Ibuprofen are responsible for hundreds of thousands of emergency room visits annually!

The aforementioned groundbreaking medical discoveries are currently being utilized with the use of marijuana extracts in the form of CBD (Cannabidiol),[1] and **RS (Rick Simpson)[2] oils.

Deaths in the U.S. Annually, Associated With the Use of the Following Substances.

Tobacco	450,000
Alcohol	150,000
Prescription Drugs	100,000
Cocaine / Heroin	15,000
Caffeine	5,200
Aspirin	1,000
Marijuana	0

[11]

---

1   Derived from hemp (non THC)
2   Derived from organic cannabis (Contains THC): Much more therapeutic than CBD's when dealing with catastrophic illness.

Our founding fathers intended freedom to be based on a doctrine of "Usufruct," (life, liberty, and the pursuit of happiness.) That's why George Washington grew so much hemp!

> *"Make the most of hemp seed, sow it everywhere!"*
>
> — (George Washington)

However, over the years, millions of Americans have fallen victim to the enforcement of Draconian laws which have persecuted them for their use of this innocuous herb.

> *"Casual drug users should be taken out and shot"*
>
> — (Chief Darryl Gates, head of Los Angeles Police Department in front of the United States Judiciary Committee in 1990)

Some of these people were beautiful, spiritually evolved human beings, and like the native Americans who preceded them a century before, they wanted independence and freedom from the corruption and constraints of modern civilization; to live off the land and be in harmony with nature. But instead, these so-called Flower Children were labeled as misfits, outcasts, or black sheep by the conservative, right wing majority that was predominant during this era.

People like the late Jack Herer (film maker, author, and pioneer of political hemp activism), Dennis Peron (politician, activist, and former presidential candidate), and Jerry Brown (Former Attorney General and three term Governor of California), are some of the champions and patriots of the medicinal marijuana movement, who brought true compassion and justice for one of our most precious inalienable rights: The right to use natural medicine free from persecution! [12]

HOT BURNING CANDY: POEM BY NEIL M PINE

"Hot Burning Candy
Sweet Plumes of Smoke Everywhere
A Mystery Surrounds Me
Is There Magic in the Air?
I Smoke It in the Morning
There's Glory when I Rise
I Smoke it in the Afternoon
Brings Tears to My Eyes
And After a Big Dinner
I Smoke it at Night
And Just Like an Ocean Breeze
It Makes Me Feel Alright
Rastafarian Revolution
Keep Me Free from Persecution
Smoke the Herb
Smoke the Herb
Smoke the Herb
And you'll See
How Could God Put it on This Earth
If it Wasn't Meant for Me!

5. The State of the Art Holistic Medical Equipment

a. Ozone Therapy Machine: Oxygenates the blood orally, rectally, or intravenously

b. Rife Machine: Banned by the FDA and the AMA, Developed over 50 years ago; this machine may demonstrate miraculous healing properties for many conditions. The technology is based on directing frequencies and vibrations to afflicted areas of the body.

c. Hyperbaric Chamber: Oxygenates the blood and is very effective towards cancer treatment, auto immune deficiency, and healing wounds.

d. Alpha Wave Machine: Helps a person to focus on deep states of meditation and relaxation by teaching them how to harness Alpha Waves.

e. Aroma Therapy Equipment: For relaxation and meditation.

f. Water Massage Machine: Therapeutic, relaxing and rejuvenating

g. Saline Swimming Pool & Jacuzzi (free from chlorine and other harsh chemical additives)

h. Sauna (dry or wet): Sweating builds up circulation and helps to eliminate toxins in the body [13]

6. The State of the Art Living: (Naturally and Bio-sustainably)

a. * Green Cosmetics and Toiletries: (non-petroleum or chemical based). Shampoos using tea tree oil, and other more gentile herbal ingredients. Natural mouthwashes, which contain disinfecting and astringent properties, like witch-hazel and spearmint oil. Natural toothpastes, free from the Saccharine, NutraSweet and Fluorides of the big brand names. Fluoride, even in minute dosages, over a long period of time, culminates in the central nervous system, therefore potentially causing severe neurological damage. The Nazi's put fluoride in the drinking water at their infamous concentration camps, but it wasn't to help anybody with their teeth! (See chapter 7: Pollution)

b. Natural makeup and foundation: The latest State of the Art makeup uses only minerals, doesn't clog the pores or irritate the skin. Skin moisturizers, face and body creams, using only aloe vera, cocoa butter, lanolin*, avocado, lavender and almond oils (vegan sources of tri-omega oils), and other natural emollients.

(*Lanolin is compassionately harvested from wool, however, just like silk and honey, is technically not considered a vegan product.)

Petroleum by-products such as petroleum jelly and Methyl paraben, used in commercial cosmetics, can enter your blood stream through the skin. Read the label carefully. Use natural deodorants, free from the toxic aluminum by-products and chemical additives ubiquitous only to the major brand name manufacturers. [14]

c. * Green Cleaning Supplies: (Bio-sustainable Cleaning Products). Laundry and dish water detergents, free from harsh chemicals or soaps such as Hexachlorophene, with natural cleaning and degreasing agents. Organic disinfectants which are concentrated and all purpose (can be diluted to suit a particular job). They sanitize and leave an invigorating, fresh herbal smell. However, if you wish to completely sanitize your bathrooms and kitchens 100% hypo-allergenically, with no smells or residues, then there are a variety of steam cleaning machines currently on the market that will suit all your needs. There are dry cleaners using the state of the art green technology which also use no chemicals.[15]

d. Natural Air Fresheners: Utilizing only herbal, floral, or citrus essences. Stay away from those toxic, harsh, pungently scented brand names. Also, baking soda is an effective, safe, and gentle deodorizer for your refrigerator.[16]

(*Not only are these products biodegradable, using no chemical ingredients, but they are also compassionately manufactured. No animals suffered or were exploited in any way due to the research or manufacturing of these products.)

e. Green Pest Control: The State-of-the-Art electronic vermin and pest repellent. Advertised on TV; uses radio waves, inaudible to the human ear, to actually repel unwanted bugs and rodents. Another new green technology eradicates termites through orange oil. No more tenting your house. [17]

f. Green Compassionate Clothing and Accessories: Everything you can imagine; ties, belts, suits, dresses, shoes, sandals, socks, shirts, under garments, wallets, purses, and handbags, utilizing materials produced compassionately, free from the suffering incurred by living creatures due to wearing fur or leather. Major high-end clothing and handbag manufacturers are starting to follow this trend, utilizing vinyl's, hemp, cottons and other compassionately harvested materials. [18]

g. Utilization of Alternative Energy:

We currently have many choices of green renewable energy which can break us free from our dependence on foreign oil and the big utility companies' pretentious power grid! Millions of new jobs, in the U.S. alone, are scheduled for the alternative energy market over the next 20 years.

1. *Solar: With technology, new breakthroughs in photo voltaic cell production, biodegradable (non-petroleum based), solar substrate, and the latest advancements in the NASA solar space program, we will someday provide an indelible source of clean power!

2. *Wind: T-Boone Pickens called N. Dakota the Dubai of wind power generation. Wind power can be utilized in many ways. In several real estate projects mentioned in The Conscious Planet, wind generation is being incorporated into the design of buildings.

3. *Hydroelectric: Since the earth is 70% water, there should be no excuse not to take advantage of this untapped, clean, safe, and natural resource!

In the ocean, for example, the potential for power generation boggles the mind. Ocean Power Technologies, NASDAQ symbol, (OPTT), has a patented technology which can generate up to 100MW of electricity in as little as .39 square miles of ocean! They are currently involved in a 2 billion dollar project in Australia, and have just signed a major partnership agreement with Lockheed Martin, a 43 billion dollar corporation.

4. *Geothermal: Where geographically feasible, geothermal technology can create enough energy to power an entire city. [19]

> *There are Federal and state tax credits available, along with manufacturer's rebates, when purchasing any equipment used for the production of clean, safe, alternative energy. The meter starts running backwards as soon as you make the commitment to go green!

h. The State-of-the-Art Transportation: There are a plethora of new State of the Art vehicles, coming out to meet your State of the Art Lifestyle needs.

1. 100% electric vehicles: (non-hybrid) Scooters, motorcycles, cars, buses, and even semi-trucks, all running on pure electric power. (Electric power can be up to 90% cheaper than gasoline) [20]

Journalist for Jet Setting Magazine, Neil M. Pine, sporting a 2018 Tesla with his 20-year-old dog, Lucky!

2. Compressed Air Power: This new technology will take the country by storm. Currently, Ta Ta Motors, an Indian alternative fuel vehicle manufacturer, are currently selling this vehicle in India. However, what makes this car so fantastic is the incredible claim, that it can be driven nonstop for 1000 miles at a top speed of 95 MPH! How much will a gallon of gas cost after they sell 50 million of these vehicles?

Mark Cuban has already taken 5 million dollars directly out of his own pocket, live on the Shark Tank show, for North American distribution rights for this amazing car!

Ta Ta Motors, Air Powered Car [21]

GOT WATER?

There are kits available which claim to convert your car to partially run on straight tap water. Hydrogen fuel cells are also water, but while some people feel that Hydrogen fuel cells represent an industrial scam to manipulate the price of an almost free or relatively inexpensive commodity, others believe that it has a viable future as prototypes of this technology have already been developed and are currently in production. (See chapter 19: Hydrogen, Not Just Another Government Scam)

> 3. Ethanol & Bio-Fuel? There's new information which substantiates that bio fuels are nothing more than an enormous bureaucratic tax scam. There are too many draw backs that making these alternative fuels much less appealing than their aforementioned counterparts. One being that it still patronizes an antiquated technology (internal combustion engines). Another reason is that the manipulation of the price of certain crops, such as corn for ethanol, which when traded for fuel value, may artificially inflate the cost of commodities for food. In addition, there is evidence that biofuels, like their petro-chemical counterparts, also significantly contribute to air pollution. (Refer to chapter 18: Ethanol, Unviable Government Subsidized Tax Scam)

> i. The State of the Art Off the Grid Living
> Living off the grid, not only are you protected from EMP Attacks, Solar Flares, Black Outs and Rate Hikes, but if you harness the current State of the Art technology correctly, then you may NEVER have to pay for natural gas, gasoline, water or electricity, ever again! Advancements in photovoltaic and other alternative energy technologies is making off the grid living more appealing than ever! Furthermore, you can grow all the free food you want with all the free water you need from your own well. [23]

1. Land Conservation: Bio-sustainable Landscaping: Try using only local flora, (shrubs, trees and flowers which are indigenous to your particular climate). E.g. plant cactus with a rock garden. The average lawn is full of chemicals and takes up way too much water. You could save thousands of gallons over the years.

2. Organic Gardening: Create your own personal organic garden. This is the freshest, healthiest, and most bio- sustainable method in order for you to obtain your sustenance. (Refer to Veganic Gardening sub chapter in chapter 17: Organic vs. Conventional Foods)

3. Septic System: Contingent to living off the grid is having a State of the Art septic system. [24]
   a. Cost: In the Midwest, where the cost of living is cheaper, a septic system will run from $2,000 to $5,000, but in more affluent areas where the labor and material costs may be higher, a septic tank could go from $4,000 to as much as $12,000.

4. Drilling a Well: Before selecting an unimproved piece of land, first acquire the expertise of a licensed geologist. Not only should a water source be identified, but it also should be tested for pollution (sediment and inorganic materials). Hopefully, you will not have to drill more than 200 feet, but in some cases, people have had to drill up to 800 feet or more, which could be very costly.

   a. Cost: The cost is roughly $15-$30, per foot, and possibly up to $50 per foot depending on the terrain and rock formations.[25]

5. Water Conservation: Besides practicing organic gardening and bio-sustainable landscaping, hypothetically, just by becoming a vegan, each person can save up to one million gallons of water a year! This statistic pertains to people who eat one pound of meat a day. According to Earth Save Organization, it takes 2500 gallons of water to produce just one pound of meat as opposed to only 6 gallons of water to grow one pound of broccoli. As of 2018, The Sierra Nevada Mountains were experiencing the worst drought in over 100 years! Snow levels had dramatically declined, threatening ski resorts and regional farming interests.

   Also, always turn off the sink while washing dishes or brushing your teeth. Another thing you can do is to collect water in a bucket while waiting for your bath water to heat up. In addition, you should replace washers in leaky sinks.

Other water saving ideas: The latest water free, foaming, bio-degradable, composting toilets. Low flow gravity toilets, tank-less water heaters, low pressure shower heads, and rain barrels for watering supplementation. Sweep your driveway and walkways, never wash them down! (Refer to chapter 8: Dust, Drought, and Desertification)

j. The State-of-the-Art Energy Conservation: High tech, energy efficient light bulbs, insulation and building materials. Energy saving thermostats, dual and triple pane, energy efficient windows, solar windows, skylights, sunrooms or solariums and atriums. Green roofing (creates a natural habitat on your roof). Green building materials, preferably hemp, bamboo, or other organic materials, which can be replenished or harvested much more efficiently than traditional lumber. Electric hedge[26]

1. Recycling:

a. (Paper, plastic, glass, copper, and aluminum) [27]

b. Official E-waste disposal site: The Green Spot(Batteries, TV's, Computers, and other discarded electronic items) [28]

ADVANTAGES OF OFF THE GRID LIVING:

No Electric Bill: Photo Voltaic technology. Solar panels on the roof can charge a solar battery in your home. In addition, a home wind generation system can also be very economical back up system.

No Natural Gas Bill: PG&E is being sued and will inevitably pass down these costs. As time goes by (with rising utility costs and the advancements of technology), the viability of off the grid living can only ameliorate. Now is the time to get away from all these pretentious utility companies!

No Gasoline Cost: Never go to the gas station again. The solar battery in your home can completely charge your high range electric vehicle.

No Water Bill: After initial startup cost in the drilling and creation of a well, you should thereafter, be able to get all the water you need to cook, wash clothes, bathe and for gardening.

Protection from EMP attacks (Electro Magnetic Pulse): If a rogue nation, such as North Korea or Iran, were to detonate a small nuclear device high in the atmosphere, then this could potentially take down a significant portion of our Grid in the U.S. [29]

Protection from Solar Flares: In the same respect, a solar flare could also wreak havoc with the grid. (Refer to the Carrington event of 1859) [30]

Protection From Rate Hikes: S.C.E. (Southern California Edison), and PG&E (Pacific Gas and Electric), are facing severe law suits from the egregious fires that took place in 2018, in California, and will inevitably be passing these costs down to consumers.[31]

Protection from Black Outs: Rolling blackouts could be initiated during peak hours in many parts of the country due to extreme weather conditions or some of the other aforementioned disasters.

More Availability and Options: Choosing an off the grid building site potentially affords much more value, options and versatility in selecting the perfect property. It's easy and cheap to find raw, unimproved land away from all utilities.

> *"The elimination of all antediluvian energy technologies (coal, oil, and nuclear), and the development of truly sustainable energy sources (wind, solar, and hydroelectric), will someday represent a monumental epoch in the evolution of humanity! — (Neil M. Pine)*

Notes:

[1]   https://www.greenamerica.org/blog/dangers-genetically-modified-foods- guardian

[2]   https://www.cancer.gov/about-cancer/causes-prevention/risk/diet/artificial-   sweeteners-fact-sheet

[3]   https://www.medicinenet.com/macrobiotic_diet/views.htm

[4]   https://r.search.yahoo.com/_ylt=Awr9BNZTBoJhbvsAMGdXNyoA;_ylu=Y29sbwN ncTEEc-G9zAzMEdnRpZAMEc2VjA3Ny/RV=2/RE=1635940052/RO=10/RU=https%3a%2f%2famericanaddictioncenters.org%2fover-the-counter- medications/RK=2/RS=SCs5FE8aqB7aaAKx9Owg5hXvUhE-

[5]   https://r.search.yahoo.com/_ylt=Awr9DWjnBoJhOKwArlBXNyoA;_ylu=Y29sbwN   ncT-EEcG9zAzIEdnRpZAMEc2VjA3Ny/RV=2/RE=1635940199/RO=10/RU=https%3a%2f%2fwww.mcgill.ca%2foss%2farticle%2fhealth-you-asked%2ffood-   irradiation-dangerous/RK=2/RS=JKc0ED4ARsjGnkHJyiEHbhlxV9k-

[6]   https://r.search.yahoo.com/_ylt=Awr9DRVKC4Jh.uAAxQpXNyoA;_ylu=Y29sbwN   ncTEEc-G9zAzQEdnRpZAMEc2VjA3Ny/RV=2/RE=1635941322/RO=10/RU=https%3a%2f%2fbodyecology.com%2farticles%2fmicrowave_dangers- php%2f/RK=2/RS=igS.5QNvY5tHppCRUX1MUmNL3SA-

[7]   https://r.search.yahoo.com/_ylt=Awr9ImkTB4JhH2oABk9XNyoA;_ylu=Y29sbwNn   cTEEc-G9zAzIEdnRpZAMEc2VjA3Ny/RV=2/RE=1635940244/RO=10/RU=https%   3a%2f%2fwww.healthline.com%2fhealth%2ffood-nutrition%2fwheatgrass- benefits/RK=2/RS=fZyxesSjHDmwpZNNfE0OFlWTw3l-

[8]   https://r.search.yahoo.com/_ylt=Awr9lmh5DYJhHvYA2NZXNyoA;_ylu=Y29sbwN   ncT-
EEcG9zAzEEdnRpZAMEc2VjA3Nj/RV=2/RE=1635941881/RO=10/RU=https%3a%2f%2fplant-based.
org%2fwhich-supplements-are-necessary-for-   vegans%2f%23%3a~%3atext%3dWhich%2520sup-
plements%2520are%2520neces   sary%2520for%2520vegans%2520The%2520only%2cin%-
2520cell%2520metabolis   m%2520and%2520red%2520blood%2520cell%2520production./RK=2/
RS=99.zdJ Bq8UxeFmTU3BJRBA9EzlY-

[9]   https://r.search.yahoo.com/_ylt=Awr9DszEDYJhYagANmlXNyoA;_ylu=Y29sbwNn   cTEEc-
G9zAzUEdnRpZAMEc2VjA3Ny/RV=2/RE=1635941956/RO=10/RU=https% 3a%2f%2fwww.nbcnews.
com%2fbetter%2fhealth%2fbest-way-do-summer-detox- according-nutritionists-ncna885161/RK=2/
RS=fGwHjyIiisdfCp3CO2uWAaXj24Y-

[10]   https://r.search.yahoo.com/_ylt=Awr9lmEjDoJheacAa11XNyoA;_ylu=Y29sbwNncT EEcG-
9zAzEEdnRpZAMEc2VjA3Ny/RV=2/RE=1635942051/RO=10/RU=https%3a%2f%2fwww.thehealth-
board.com%2fwhat-are-the-different-types-of-holistic-   practitioners.htm/RK=2/RS=2u9zMElSCOc.
AqYJ7n5T8z0JnEU-

[11]   https://r.search.yahoo.com/_ylt=AwrgDumXDoJh1S4ABg5XNyoA;_ylu=Y29sb   wNncT-
EEcG9zAzMEdnRpZAMEc2VjA3Ny/RV=2/RE=1635942168/RO=10/RU=https%3a%2f%2fwww.mpp.
org%2fspecial%2fmarijuana-is- safer%2f/RK=2/RS=k9kDJ9zSyP0R6PDpJvxKluhFMMQ-

[12]   https://r.search.yahoo.com/_ylt=Awr9DsopEoJhamkAFIlXNyoA;_ylu=Y29sbwNncTEEcG9zAzIEd-
nRpZAMEc2VjA3Ny/RV=2/RE=1635943081/RO=10/RU=https%3a%2f%2fbigbudsmag.com%2fjack-her-
er-marijuana-activist-marijuana- strain%2f/RK=2/RS=8eMwGwTVcfGSMb.mrGWiSqZDSZ8-

[13]   https://r.search.yahoo.com/_ylt=Awr9IlDZDoJhicgAAjVXNyoA;_ylu=Y29sbwNncT EEcG-
9zAzIEdnRpZAMEc2VjA3Ny/RV=2/RE=1635942234/RO=10/RU=https%3a%2f%2fwww.frequencyris-
ing.com%2f/RK=2/RS=OLPqVr4n3Mz34X0z_nkTEtKh vBU-

[14]   https://r.search.yahoo.com/_ylt=Awr9JhcWD4JhO3YAvhJXNyoA;_ylu=Y29sbwNn cTEEcG-
9zAzEEdnRpZAMEc2VjA3Ny/RV=2/RE=1635942294/RO=10/RU=https% 3a%2f%2fwww.acme-hard-
esty.com%2fgreen-cosmetics-sustainable- beauty%2f/RK=2/RS=npfY8x7RP056CLENk6aydG8LCbl-

[15]   https://r.search.yahoo.com/_ylt=Awr9BNk9D4JhD2lAfiNXNyoA;_ylu=Y29sbwNnc TEEcG-
9zAzQEdnRpZAMEc2VjA3Ny/RV=2/RE=1635942334/RO=10/RU=https%3a%2f%2fwww.thespruce.
com%2fbest-green-cleaning-products- 4776786/RK=2/RS=nYDMo.m0LfNL2PDa7Opx6jir_pc-

[16]   https://r.search.yahoo.com/_ylt=Awr9Ikx6D4Jh0zcAXwVXNyoA;_ylu=Y29sbwNnc TEEcG-
9zAzMEdnRpZAMEc2VjA3Ny/RV=2/RE=1635942394/RO=10/RU=https%3a%2f%2fwww.myairfresh-
ener.com%2fmygreen/RK=2/RS=X08OFBE9wxEuQ9Gi ULX1elebUOc-

[17]   https://r.search.yahoo.com/_ylt=Awr9Ds61D4JhoKIAICdXNyoA;_ylu=Y29sbwNnc   TEEc-
G9zAzEEdnRpZAMEc2VjA3Nj/RV=2/RE=1635942453/RO=10/RU=https%3a%2f%2fwww.gee-
kwrapped.com%2fguides%2fbest-electronic-pest-control-   devices%23%3a~%3atext%3dElectro-
magnetic%2520pest%2520control%2520uses%2520existing%2520wiring%2520in%2520the%2cof
%25202017%2520haven%25 E2%2580%2599t%2520found%2520evidence%2520that%2520it%252
0helps./RK= 2/RS=4zPpplXZatZBefi8CcQxCpb5PQk-

[18]   https://r.search.yahoo.com/_ylt=Awr9IlfkD4JhS0MANsJXNyoA;_ylu=Y29sbwNnc   TEEcG-
9zAzQEdnRpZAMEc2VjA3Ny/RV=2/RE=1635942501/RO=10/RU=https%3a%2f%2fwww.thegood-
trade.com%2ffeatures%2feco-friendly-clothing- brands/RK=2/RS=aw1TVyNBnahpInQGjiVYI4SiyKA-

[19]   https://r.search.yahoo.com/_ylt=Awr9DskNEIJh01oACU1XNyoA;_ylu=Y29sbwNnc TEEc-
G9zAzIEdnRpZAMEc2VjA3Ny/RV=2/RE=1635942542/RO=10/RU=https%3a%2f%2fwww.altenergy.
org%2f/RK=2/RS=xPBTvAIANJYc0bbDYEKmDrd4tJgM-

[20]   https://r.search.yahoo.com/_ylt=AwrgDuccE4Jhue8AwENXNyoA;_ylu=Y29sbwNnc TEEc-
G9zAzEEdnRpZAMEc2VjA3Ny/RV=2/RE=1635943325/RO=10/RU=https%3a%2f%2fwww.edmunds.
com%2felectric- car%2f/RK=2/RS=L4MzCnIP64H4OJ5WjuyalB997eA-

[21]   https://en.wikipedia.org/wiki/Compressed_air_car

[22]   https://automobiles.honda.com/clarity-fuel-cell

[23]   https://r.search.yahoo.com/_ylt=Awr9Duy4FoJh0fMAERJXNyoA;_ylu=Y29sbwNn cTEEcG-
9zAzQEdnRpZAMEc2VjA3Ny/RV=2/RE=1635944248/RO=10/RU=https% 3a%2f%2fwww.greenand-

growing.org%2fhow-to-live-off-the- grid%2f/RK=2/RS=ZaEgBE0c9GIJda.mx7a6amKDriU-

[24]   https://www.epa.gov/septic/advanced-technology-onsite-treatment- wastewater-products-approved-state

[25]   https://home.costhelper.com/well-drilling.html

[26]   https://r.search.yahoo.com/_ylt=Awr9Fq0kGoJhenMAXXNXNyoA;_ylu=Y29sbwN ncTEEc-G9zAzMEdnRpZAMEc2VjA3Ny/RV=2/RE=1635945124/RO=10/RU=http%3a%2f%2fwww.americanli-metechnology.com%2fwhat-is- hempcrete%2f/RK=2/RS=6.NBf0QnAreuqmR31dUSkkrdzBw-

[27]   https://www.facebook.com/WarOnWaste.WOW/

[28]   https://www.manta.com/c/mhk10ws/the-green-spot-recycling-corporation

[29]   https://en.wikipedia.org/wiki/Nuclear_electromagnetic_pulse

[30]   https://en.wikipedia.org/wiki/Solar_storm_of_1859

[31]   https://www.theguardian.com/us-news/2018/nov/16/california-wildfires-causes- investigation-pge

Live The State of the Art Lifestyle

and become a part of …

The Conscious Planet

List of Photos :

List of Cartoons

List of Illustrations:

Professional Endorsements:

1.) "Neil has always been a passionate advocate for animal rights, environmental awareness and health consciousness! This book addresses the human imperative of sustainability while living in harmony with nature and teaching compassion!" (Mark Thompson)

Fox TV Los Angeles
American Idol
TYT (The Young Turks)
KFI Radio Los Angeles
KGO Radio San Francisco
Meteorologist

2.) "Author Neil M. Pine is a true pioneer of the Vegan movement! For over 40 years he has been a champion for animal rights, environmental awareness, and holistic health concepts! We need a book like his now, more than at any other time in history!" (Claudia DeSantis)

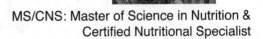

MS/CNS: Master of Science in Nutrition &
Certified Nutritional Specialist

3.) "A Must read if you care about your health and the future of our children!" (Dr. Zayd Ratansi ND)

Managing Director of …

4.) "A Game Changer! This book challenges many core ideas and values related to parenting choices. The author provides an alternative perspective for parents to evaluate the relationships between food choices and parenting practices that may have a profound impact on the child." (Dr. Carol Sigala, PhD)

Professor of Child Development

5.) "Ethical, environmental, humanitarian, & healthy veganism could be the quickest way to change the world" — (Lama Jigme Gyatso)

The rational contemplative and Spiritual Directive of Buddha Joy Meditation School

6.) "Ground breaking cutting edge journalism which is not only informative and compassionate, but is also critical toward the survival of the planet" (Mike Russell)

Mike Russell

Publisher for Jet Setting Magazine

7.) "Holistic, Anthropogenic and Compassionate. The Future of Humanity!" (Carmela Evangelista)

ETM: Evangelista Talent Management

Beverly Hills Hotel (1983)

Former 3-time California Governor, Jerry Brown, with author Neil M. Pine at a $1000.00 a plate fund raiser dinner event, at the Beverly Hills Hotel in 1983. Mr. Pine enjoyed an exquisitely prepared *Vegan* meal while discussing *Vegan* and compassionate/420 politics (13 years prior to the passing of Prop. 215) Brown was instrumental in the passing of this Bill in 1996.

- In loving memory of Lucky: Miniature pincher. He lived till 21 years old!

(Illustrations by Neil M Pine)

(Lucky with his mom, Brenda Yoon, and the late Florence Pine (the authors mother), 2017)

- https://blog.jetsettingmagazine.com/news/miracle-dog-going-on-21-years-old

Index

Symbols

9-11 30, 76, 80, 82, 83, 85, 86, 88
60 Minutes 25, 87
400 Chernobyl's 245

A

AA 190, 194
ABC Nightly News 108
Acid Rain 117, 124
Adams, Carol 38, 39, 60
adipose products 17, 123
Africa 66, 146, 158, 159, 252, 276, 277,
    282, 283, 291-293, 296
African Wild Dog 282
agave 306
Age of Turbulence, The 82
Agribusiness Council 232
AHIMSA 8, 16. 39
Air Powered Car 316
Akers, Keith 53, 54, 61
Alberswerth, David 92
ALCOA 125, 127, 128, 129
Alcoholism: The Cause & The Cure 195
alternative energy 12, 73, 77, 78, 93, 95,
    98, 102-106, 234, 237, 241, 242,
    304, 315, 317
Alzheimer's disease 120, 177, 203
Amazonia 155
Amazon rainforest 149
American Bison Society 282
American Burying Beetle 275
American Civil Liberties Union (ACLU)
    81
American Dental Association (ADA) 131
American diet 201
American-Made Energy and Good Jobs
    Act 90

American Medical Association (AMA)
    11, 201, 313
American Petroleum Institute 89
American Psychiatric Association 27
American wilderness 18
ammonia 20, 97, 139
Ammonium Hydroxide ("Pink Slime")
    12, 174
Amos 51, 60
Amphibians 51, 60
Amstrup, Steve 278
animal abuse 27
Animal Liberation Front (ALF) 68
animal rights 15, 51, 52, 62, 67-69
animal sacrifice 53, 54
Animal slaughter 53, 54
Anthropocentrism 26, 41, 59
Anthropogenic User of Land 148
Anthropology 10
antibiotics 107, 149, 172, 173, 178, 201-
    203, 205, 206, 212, 226, 306
Antimony 120
Apocalyptic fears 47
*Approaching Hoof Beats: The Four Horse-
    men of the Apocalypse* 46
Aspartame 221, 306
Attention Deficit Disorder (ADD) 122,
    188, 189, 213
Autry, Gene 23
Aventis 221, 223, 224

B

Baboumian, Patrik 71
Baer, Robert 80
Baker, Russell 32, 59
Banks, Tyra 63
Barnard, Neal 33, 68, 70, 71, 203
Baskin-Robins 70
Bayer 108, 217, 219, 221, 223, 224
Becker, William 232
Beck, Glen 88
Beecher, Henry Ward 302